Dimensions Ma
Tests 5A

Author

Dawn Yuen

Singapore Math Inc.

Published by Singapore Math Inc.

19535 SW 129th Avenue
Tualatin, OR 97062
www.singaporemath.com

Dimensions Math® Tests 5A
ISBN 978-1-947226-55-5

First published 2020
Reprinted 2020, 2021, 2022 (twice), 2023, 2024

Printed in China

Acknowledgments

Design and illustration by Cameron Wray with Carli Bartlett.

Preface

Dimensions Math® Tests is a series of assessments to help teachers systematically evaluate student progress. The tests align with the content of Dimensions Math K–5 textbooks.

Dimensions Math Tests K uses pictorially engaging questions to test student ability to grasp key concepts through various methods including circling, matching, coloring, drawing, and writing numbers.

Dimensions Math Tests 1–5 have differentiated assessments. Tests consist of multiple-choice questions that assess comprehension of key concepts, and free response questions for students to demonstrate their problem-solving skills.

Test A focuses on key concepts and fundamental problem-solving skills.

Test B focuses on the application of analytical skills, thinking skills, and heuristics.

Contents

25 min **Score**

40

Test A

Chapter 1 Whole Numbers

Section A (2 points each)
Circle the correct option: **A**, **B**, **C**, or **D**.

1 The digit 1 in 51,482,437 stands for _______________.

A 1 one

B 1 million

C 1 thousand

D 1 hundred

2 Seven hundred thirty-five million, two hundred ten is _______________.

A 700,350, 210

B 735,210,000

C 735,000,210

D 735,210

3 How many ten thousands are there in 280,000?

    A 280   **B** 2,800

    C 28,000   **D** 28

4 $80 \times 7{,}000 =$ ☐

    A 56,000   **B** 560,000

    C 5,600   **D** 5,600,000

5 $360{,}000 \div 4{,}000 =$ ☐

    A 90   **B** 9,000

    C 90,000   **D** 900

Section B (2 points each)

6 Write the number 10,905,122 in words.

7 Arrange the following digits to form the least seven-digit number.

0, 3, 5, 8, 7, 4, 1

8 Write the missing number.

$20{,}138{,}085 = 20{,}000{,}000 + \underline{\hspace{3cm}} + 30{,}000 + 8{,}000 + 85$

9 One million less than 584,920,119 is \underline{\hspace{3cm}}.

10 Use the given letters to put the numbers in order from least to greatest.

750,356,815	705,356,815	750,536,815	750,356,851
A	B	C	D

11 Write >, <, or = in the ◯.

Twenty hundred thousands ◯ 20 millions

12 32,000 × 200 = ☐

13 57 × ☐ = 570,000

14 $30 \times 100 \times 10 =$ ☐

15 $4{,}300{,}000 \div 1{,}000 =$ ☐

16 $3{,}500 \times 40 =$ ☐

17 $180{,}000 \div 60 =$ ☐

18 $49{,}000 \div 700 =$ ☐

19 A popcorn machine costs \$14,000. How much does it cost a movie theater chain to buy 20 popcorn machines?

20 10,000 balloons are packed in bags of 50 balloons each. How many bags of balloons are there?

25 min   **Score**

40

Test B

Chapter 1 Whole Numbers

Section A (2 points each)
Circle the correct option: **A**, **B**, **C**, or **D**.

1 Which digit in 130,215,640 is in the ten millions place?

A 3

B 5

C 2

D 6

2 The value of the digit 2 in the number 32,785,416 is _______________.

A 20,000,000

B 200,000

C 2,000,000

D 2,000

 Ten hundred thousands more than 789,654,321 is _______________.

A 789,754,321

B 790,654,321

C 789,664,321

D 780,654,421

4 12,600,000 is _______________ times as great as 126,000.

A 1,000

B 10

C 10,000

D 100

5 320,000,000 ÷ 80 = [ ]

A 400,000

B 4,000,000

C 40,000

D 40,000,000

Section B (2 points each)

6 Write the following number.

Five hundred eighty-three million, ten thousand, eleven

7 Arrange the digits 0, 0, 1, 2, 5, 6, 9 to form the greatest seven-digit odd number.

8 Write the missing number.

$18{,}206{,}551 = 551 +$ ⬚ $+ 18{,}000{,}000$

9 842,700,000 has ________________ hundred thousands.

10 Write the following in order from greatest to least.

50 hundred thousands, five hundred millions, 5 ten millions

11 What is the greatest even number that is less than $15{,}000{,}000 + 90{,}000 + 333$?

12 $10{,}000 \times 530 =$ ☐

13 $8{,}000 \times 5{,}000 =$ ☐

14 $\boxed{} \times 28 = 28{,}000{,}000$

15 $990{,}191{,}011 = 1{,}000 + 900{,}000{,}000 + \boxed{} + 11 + 190{,}000$

16 $10{,}000 \times 40 \times 10 = \boxed{}$

17 $7{,}180{,}000 \div 100 = \boxed{}$

18 Write >, <, or = in the ◯.

$25{,}000 \times 300 \,\bigcirc\, 3{,}000 \times 2{,}500$

19 A factory packed 600,000 envelopes equally into 5,000 boxes. How many envelopes are in each box?

20 A hotel spent $4,000 renovating each guest room. It renovated 21 guest rooms. How much money did the hotel spend on renovating guest rooms altogether?

Name: ________________________________

Date: ________________________________

40

Chapter 2 Writing and Evaluating Expressions

Section A (2 points each)

Circle the correct option: **A**, **B**, **C**, or **D**.

1 $55 - (10 - 5) =$ ⬚

A 40   **B** 50

C 45   **D** 5

2 $90 \div (3 + 27) =$ ⬚

A 3   **B** 57

C 30   **D** 120

3 $4 \times 25 + 3 \times 12 =$ ☐

A 1,236

B 1,344

C 136

D 340

4 Which of the following expressions has the same value as $20 + 10 - 12 \div 2 \times 3$?

A $20 + 10 - 12 \div (2 \times 3)$

B $(20 + 10) - (12 \div 2) \times 3$

C $(20 + 10 - 12) \div 2 \times 3$

D $20 + (10 - 12 \div 2) \times 3$

5 A baker made 30 muffins yesterday. He sold 21 muffins at \$3 each in the morning and the rest of the muffins at \$2 each in the afternoon. Which expression shows the total amount of money in dollars he received from selling the muffins yesterday?

A $21 \times 3 + 30 - 21 \times 2$

B $(21 + 13) \times (3 + 2)$

C $21 \times 3 + 30 \times 2$

D $(21 \times 3) + (30 - 21) \times 2$

Section B (2 points each)

6 $36 \div (9 \times 2) = \boxed{}$

7 $90 - 640 \div 8 + 10 = \boxed{}$

8 $8 \times (20 - 5 \times 2) = \boxed{}$

9 $2 \times (60 \div 5) \times (8 - 3) = \boxed{}$

10 $(9 + 4) \times 12 = \boxed{} \times 12 + 4 \times \boxed{}$

11 $39 \times 22 + 39 \times 7 = 39 \times \boxed{}$

12 $998 \times 67 = (1{,}000 - \boxed{}) \times 67$

13 Write $>$, $<$, or $=$ in the $\bigcirc$.

$5 \times (50 - 5) \bigcirc 50 - 5 \times 5$

14 Insert ( ) to make the equation true.

$10 \times 3 + 7 = 100$

15 Laila had \$55. She bought a book for \$7, 3 cards for \$2 each, and a concert ticket for \$12. How much money does she have left?

16 There are 100 cartons of eggs with 12 eggs in each carton and 75 cartons of eggs with 6 eggs in each carton. 15 of the eggs broke. What is the total number of eggs left?

 A child train ticket costs half the price of a regular ticket. A senior ticket costs $10 less than a regular ticket. Dion's mom paid $230 for three regular tickets, two child tickets, and one senior ticket. What is the price of a regular train ticket?

Name: ______________________

Date: ______________________

40

Test B

Chapter 2 Writing and Evaluating Expressions

Section A (2 points each)

Circle the correct option: **A**, **B**, **C**, or **D**.

1 $20{,}000 \div (200 \div 20) = \boxed{}$

A 20	**B** 200
C 2,000	**D** 100

2 $12 + 16 \div 4 - 3 = \boxed{}$

A 4	**B** 11
C 28	**D** 13

3 $50 - (8 + 2) \times (16 - 12) =$ [ ]

A 160

B 628

C 10

D 692

4 Which expression does NOT have the same value as $(5 \times 4) + 6 - (9 \div 3)$?

A $5 \times 4 + 6 - (9 \div 3)$

B $5 \times (4 + 6 - 9 \div 3)$

C $(5 \times 4 + 6) - 9 \div 3$

D $5 \times 4 + 6 - 9 \div 3$

5 A party center charged \$215 per party for up to 15 children, and \$4 per additional child. Which expression shows the total cost of a party for 25 children?

A $215 + 4 \times 25 - 15$

B $(215 \div 15) + (4 \times 10)$

C $215 + 4 \times 25$

D $215 + 4 \times (25 - 15)$

Section B (2 points each)

6 $180{,}000 \div (60{,}000 - 30{,}000) =$ ☐

7 $3 + 3 + (3 \div 3 \times 3) - 3 \times 3 =$ ☐

8 $12 \times 2 - 24 \div 3 \div 2 =$ ☐

9 $(5 - 1) \times 3 \times (6 \div 2 + 7) + 10 \div 2 =$ ☐

10 $900 \times (100 - 75) = 900 \times$ ☐ $-$ ☐ $\times 75$

11 $10{,}000 \times 7{,}000 - 500 \times 10{,}000 = 10{,}000 \times \boxed{}$

12 $124 \times 69{,}997 = \boxed{} \times 70{,}000 - 124 \times \boxed{}$

13 Write >, <, or = in the $\bigcirc$.

$(85 - 50) \times 10 + 2 \bigcirc 2 + 10 \times (70 \div 2)$

14 Insert ( ) to make the equation true.

$6 + 4 \times 10 \div 2 = 50$

Section C (4 points each)

15 Kaylen's neighbors went on a vacation for 10 days. They paid Kaylen $2 a day to walk their dog when they were on vacation. They paid her twice as much per day to take care of their cats. They also paid her $12 for watering their plants for the whole time that they were gone. How much did they pay Kaylen per day?

16 There are 50 more than twice as many red balls as green balls in a ball pit. There are half as many yellow balls in the pit as red balls. There are 105 yellow balls in the pit. What is the total number of red, green, and yellow balls in the pit?

 Sofia had twice as many stickers as Mei. After Mei used up one third of her stickers and Sofia gave her 10 stickers, Sofia still had twice as many stickers as Mei. How many stickers did they have altogether at first?

Test A

Chapter 3 Multiplication and Division

Section A (2 points each)

Circle the correct option: **A**, **B**, **C**, or **D**.

1 $51 \times 78 = $ [____]

A 3,578   **B** 3,978

C 3,618   **D** 129

2 $3,105 \times 51 = $ [____]

A 158,355   **B** 15,355

C 3,156   **D** 158,155

3 $510 \div 70$ is _______.

A 7

B 440

C 7 R 20

D 72 R 6

4 $(500 + 250) \div 25 =$ [ ]

A 510

B 750

C 20

D 30

5 How many eggs are needed to fill 240 cartons that can hold 12 eggs each?

A 20

B 2,880

C 288

D 2,400

6 $209 \times 55 =$

7 $4{,}522 \times 36 =$

8 $14 \times 125 \times 6 =$

9 Divide 70 by 20.

10 Divide 59 by 12.

11 Divide 831 by 29.

12 Divide 7,200 by 75.

13 Write >, <, or = in the ◯.

398 × 52 ◯ 14,885 + 2,988

14 Express 2,645 minutes as hours and minutes.

Section C (4 points each)

15 A baby seal weighs 15 kg. A baby whale weighs 181 times as much as the baby seal. How much more does the baby whale weigh than the baby seal in kilograms?

16 A teacher distributed 956 counters equally among 24 students. How many counters did each student get and how many were left over?

 A concert hall sold 95 regular tickets at the same price per ticket for a total of $3,990. It also sold 38 discounted tickets for half the price of a regular ticket. How much money did it receive altogether from the sale of the discounted tickets?

Test B

Chapter 3 Multiplication and Division

Section A (2 points each)

Circle the correct option: **A**, **B**, **C**, or **D**.

1 $807 \times 16 =$ ☐

   A 12,872   **B** 50

   C 12,912   **D** 8,592

2 $7{,}455 \times 97 =$ ☐

   A 686,405   **B** 7,522

   C 72,313   **D** 723,135

3 8,808 ÷ 40 is _______.

A 220

B 2,202

C 220 R 8

D 22 R 8

4 2 × (120 + 480) ÷ 12 = ☐

A 280

B 100

C 60

D 50

5 What is the area of a soccer field that is 105 m long and 68 m wide?

A 346 m²

B 173 m²

C 7,140 m²

D 6,140 m²

Section B (2 points each)

6 95 × 319 =

7 48 × 8,073 =

8 41 × 111 × 11 =

9 Divide 95 by 26.

10 Divide 72,060 by 60.

11 Divide 359 by 72.

12 Divide 1,197 by 52.

13 Write >, <, or = in the ◯.

49 × 29 ◯ 69,550 ÷ 73

14 Express 4,810 inches as feet and inches.

Section C (4 points each)

15 A tiger weighs 685 lb. An orangutan weighs 208 lb. An African elephant weighs 14 times as much as the tiger and orangutan combined. How many pounds does the elephant weigh?

16 A factory produced 500 lb of cat treats last week and these treats were packed in containers that hold 21 oz of treats each. How many containers did the factory fill and how many pounds and ounces of cat treats were left unpacked?

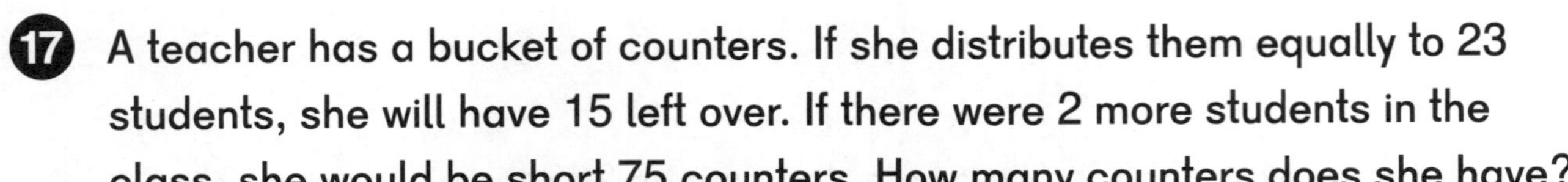

17 A teacher has a bucket of counters. If she distributes them equally to 23 students, she will have 15 left over. If there were 2 more students in the class, she would be short 75 counters. How many counters does she have?

Name: ___________________________________

Date: ___________________________________

Test A

Chapter 4 Addition and Subtraction of Fractions

Section A (2 points each)

Circle the correct option: **A**, **B**, **C**, or **D**.

1 $\frac{1}{5} + \frac{3}{4} =$ ☐

   A $\frac{4}{20}$ **B** $\frac{4}{9}$

   C $\frac{19}{20}$ **D** $\frac{3}{20}$

2 $2\frac{1}{9} + 1\frac{2}{3} =$ ☐

   A $3\frac{7}{9}$ **B** $3\frac{3}{9}$

   C $2\frac{7}{9}$ **D** $3\frac{2}{3}$

3 $\frac{11}{12} - \frac{1}{2} - \frac{1}{6} =$

    A $\frac{5}{12}$ **B** $\frac{3}{4}$

    C $\frac{1}{2}$ **D** $\frac{1}{4}$

4 $4\frac{3}{4} - \frac{1}{7} =$

    A $4\frac{1}{2}$ **B** $4\frac{2}{7}$

    C $4\frac{17}{28}$ **D** $4\frac{25}{28}$

5 $4\frac{7}{16}$ lb is _______ lb heavier than 2 lb.

    A $6\frac{7}{16}$ **B** $\frac{7}{16}$

    C $2\frac{9}{16}$ **D** $2\frac{7}{16}$

Section B (2 points each)

Express all the answers in simplest form.

6 Find the value of $130 \div 4$.

7 Add $\frac{7}{3}$ and $\frac{3}{2}$.

8 $4\frac{2}{5} + 1\frac{2}{3} =$

9 Find the value of $8\frac{1}{4} + \frac{1}{6} + \frac{3}{4}$.

10 Subtract $\frac{2}{3}$ from $\frac{3}{4}$.

11 $2\frac{4}{7} - \frac{9}{14} =$ ☐

12   Find the value of $3\frac{7}{10} - 1\frac{1}{2}$.

13 $3\frac{1}{2} - 1\frac{2}{3} =$ ☐

14 $3\frac{1}{3} + \left(2\frac{2}{3} - 2\frac{1}{6}\right) =$ ☐

Section C (4 points each)
Express all the answers in simplest form.

15 A juice bar employee mixed 5 L of carrot juice and 9 L of orange juice and then poured the mixture into 12 bottles equally. How many liters of juice are in each bottle?

16 A movie theater used $4\frac{1}{4}$ lb of popcorn kernels on Monday to make popcorn. It used $1\frac{1}{8}$ lb more popcorn kernels on Tuesday than on Monday. How many pounds of popcorn kernels did the theater use altogether in those two days?

 Amy had 8 ft of ribbon. She used $4\frac{1}{3}$ ft of the ribbon to tie a present and $2\frac{5}{6}$ ft to make a bow. She threw away the rest of the ribbon. How many feet of ribbon did she throw away?

Name: ________________________

Date: ________________________

Chapter 4 Addition and Subtraction of Fractions

Section A (2 points each)

Circle the correct option: **A**, **B**, **C**, or **D**.

1 $\frac{1}{3} + \frac{1}{4} + \frac{1}{8} =$ [____]

 A $\frac{17}{24}$ **B** $\frac{7}{12}$

 C $\frac{8}{12}$ **D** $\frac{11}{24}$

2 $3\frac{1}{10} + 4\frac{2}{15} + 5\frac{3}{5} =$ [____]

 A $12\frac{3}{5}$ **B** $12\frac{5}{6}$

 C $\frac{5}{6}$ **D** $7\frac{7}{30}$

3 $1 - \frac{3}{4} + \frac{1}{5} = \boxed{}$

A $\frac{1}{20}$ **B** $\frac{5}{9}$

C $\frac{9}{20}$ **D** $\frac{2}{9}$

4 $6 - \frac{1}{7} - \frac{3}{7} = \boxed{}$

A $\frac{3}{7}$ **B** $5\frac{6}{7}$

C $6\frac{4}{7}$ **D** $5\frac{3}{7}$

5 $5\frac{5}{12}$ ft is _______ ft shorter than $7\frac{1}{2}$ ft.

A $2\frac{1}{12}$ **B** $12\frac{11}{12}$

C $2\frac{1}{3}$ **D** $2\frac{11}{12}$

Section B (2 points each)

Express all the answers in simplest form.

6 Find the value of $2{,}400 \div 90$.

7 Add $\frac{8}{5}$ and $\frac{5}{8}$.

8 $9\frac{3}{4} + 5\frac{3}{16} = \boxed{}$

9 $\frac{1}{12} + 5\frac{4}{5} + \frac{5}{6} = \boxed{}$

10 Subtract $\frac{17}{24}$ from $\frac{5}{6}$.

11 $1\frac{3}{11} - \frac{4}{5} = \boxed{}$

12 Find the value of $6\frac{3}{8} - 3\frac{3}{10}$.

13 $9\frac{2}{5} - 3\frac{3}{10} - \frac{4}{5} = \boxed{}$

14 Find the value.

$(8\frac{1}{9} + \frac{1}{2}) - (\frac{11}{9} + \frac{7}{9})$

Express all the answers in simplest form.

15 A cook used $1\frac{5}{16}$ lb of beans to make a soup. She used twice that amount of beans to make a chili. How many pounds of beans did she use altogether to make the soup and the chili?

16 A pitcher can hold $3\frac{8}{9}$ L of water. A bottle can hold $3\frac{1}{5}$ L less than the pitcher. What is the total capacity of the pitcher and the bottle?

17 A baker mixed together 8 cups of rice flour and 5 cups of tapioca flour and put the mixed flour into 3 containers equally. He then used $1\frac{3}{5}$ cup of mixed flour from one of the containers. How many cups of mixed flour are left in that container?

Test A

Continual Assessment 1

Section A (2 points each)
Circle the correct option: **A**, **B**, **C**, or **D**.

1 In the number 125,232,180, the digit that has a value of 20,000,000 is in the ___________ place.

   A hundred millions   **B** ten millions

   C ten thousands   **D** millions

2 35,613,728 rounded to the nearest million is ___________.

   A 36,000,000   **B** 35,000,000

   C 6,000,000   **D** 40,000,000

3 Which of the following is an odd number that is less than 340,872 − 30,794?

   A 310,079   **B** 310,076

   C 310,075   **D** 310,078

4 4,200,000 ÷ 6,000 = ☐

A 70

B 700

C 70,000

D 7,000

5 110 − 2 × 50 + 5 = ☐

A 10

B 225

C 5

D 15

6 60 ÷ (8 − 3) × 3 = ☐

A 12

B 4

C 15

D 36

7 1,560 hr = ☐ min

A 26

B 37,440

C 93,600

D 65

8 $8,925 \div 51 = \boxed{}$

A 105

B 175

C 455,175

D 8,976

9 $\frac{8}{7} + \frac{1}{2} = \boxed{}$

A $1\frac{9}{14}$

B 1

C $\frac{9}{14}$

D $1\frac{1}{14}$

10 $2\frac{3}{8} - 1\frac{1}{5} = \boxed{}$

A $1\frac{4}{5}$

B $3\frac{23}{40}$

C $1\frac{7}{40}$

D $\frac{7}{40}$

11 Write the number that is 10,000 more than ten million.

12 $70,000,000 + 500,000 + 30,000 + 8,000 + 11 =$ ☐

13 Write >, <, or = in the ◯.

Sixty-two million, eight-five thousand nine ◯ 602,500,009

14 Use the given letters to put the numbers in order from least to greatest.

267,104,023	100,105,000	276,897,980	99,999,999
A	**B**	**C**	**D**

15 Fill in the blanks.

$(2 + 9) \times 16 = (\underline{\hspace{1cm}} \times 16) + (9 \times \underline{\hspace{1cm}})$

16 Write an expression to show the number that is twice as much as the sum of 10 and 3. Then, find the value.

17 Find the value.

$18 \times 150 \times 4$

18 Find the quotient and remainder.

$1{,}725 \div 20$

19 Divide. Express the answer in simplest form.

$200 \div 6$

20 Find the value. Express the answer in simplest form.

$1\frac{1}{2} + 2\frac{5}{6} + 1\frac{2}{3}$

21 A charity had \$84,000. It set aside \$4,900 for an event and distributed the rest of the money equally to 14 projects. How much money did each project receive?

22 The capacity of a bucket is 5 times that of a bottle. The capacity of a pitcher is $\frac{1}{2}$ that of the bucket. The bottle can hold 236 mL of water. What is the total capacity of the three containers in milliliters?

23 A vendor made 468 donut holes to sell at a fair. He put 6 donut holes in each bag and sold 65 bags. How many bags of donut holes were left unsold?

24 A supermarket sold 892 cartons of eggs with 12 eggs in each carton and 168 cartons of eggs with 6 eggs in each carton last week. How many eggs did the store sell altogether?

25 Sofia biked $2\frac{1}{5}$ miles on Friday. She biked $\frac{1}{2}$ mile more on Saturday than on Friday. How many miles did she bike on both days combined? Express the answer in simplest form.

BLANK

Name: _______________________________________

Date: _______________________________________

Test B

Continual Assessment 1

Section A (2 points each)
Circle the correct option: **A**, **B**, **C**, or **D**.

1 The value of the digit in the ten millions place of 256,452,000 is ______________ times the value of the digit in the ten thousands place.

A 10

B 10,000,000

C 100

D 1,000

2 200,000,000 is ______________ ten thousands.

A 2,000

B 20,000

C 200

D 200,000

3 What is the 16th multiple of 48?

A 768

B 720

C 3

D 64

4 600,000,000 is 6,000 ___________.

A hundreds

B ten thousands

C hundred thousands

D millions

5 $250 \div 25 + 10 \times 4 =$ [ ]

A 80

B 140

C 50

D 20

6 $28 - 2 \times (6 + 2 \div 2) =$ [ ]

A 14

B 20

C 182

D 104

7 9,860 pounds and 15 ounces is ___________ ounces.

A 157,760

B 157,775

C 9,875

D 15,791

8 21,135 ÷ 22 is ___________.

A 959 R 36

B 960 R 15

C 960

D 961

9 $\frac{1}{9} + \frac{5}{7} + \frac{4}{9} =$ [ ]

A $1\frac{28}{63}$

B $1\frac{4}{9}$

C $1\frac{17}{63}$

D $\frac{10}{63}$

10 $\frac{5}{2} - (1\frac{1}{3} + \frac{1}{2}) =$ [ ]

A $\frac{11}{6}$

B $\frac{15}{6}$

C $1\frac{2}{3}$

D $\frac{2}{3}$

11 Write 51,043,708 in words.

12 530,901,908 = 8 + 900,000 + 500,000,000 + 1,000 + [____________] + 900

13 702,200,000 ÷ [____________] = 7,022

14 Complete the number pattern.

300,170,500	312,370,500	
336,770,500		361,170,500

15 Fill in the blanks.

$421 \times 5{,}008 = 421 \times \underline{\quad\quad} + \underline{\quad\quad} \times 8$

16 Write an expression to show the number that is 3 more than the product of 7 and 11. Then, find the value.

17 Find the value.

$99 \times 734 \times 41$

18 Find the quotient and remainder.

$97{,}000 \div 98$

19 Express the fraction as a mixed number in simplest form.

$\dfrac{200}{16}$

20 Fill in the blank. Use simplest form.

$\boxed{} + 1\tfrac{1}{3} = 2\tfrac{3}{5}$

Section C (4 points each)

21 A blue whale weighs 93 times as much as a beluga whale. A humpback whale weighs 23 times as much as the beluga whale. The humpback whale weighs 65,067 lb. How many pounds does the blue whale weigh?

22 A cinema charges $8.25 for general admission. It charges $2.00 less for child admission. How much do two general and three child tickets cost altogether?

23 The price of an all-day pass to an amusement park is $30. The price of a day pass to the park is $20 and a night pass is $15. Colton and his three friends want to be in the park both during the day and the night. Find the amount of money he and his three friends will save altogether if they buy all-day passes instead of day and night passes separately.

24 A bakery sold 150 blueberry muffins and 36 fewer carrot muffins than blueberry muffins yesterday. All the muffins were packed in boxes of 6 each and sold for $11 per box. How much money did the bakery receive from selling the muffins yesterday?

25 Dion spent $\frac{2}{5}$ of his money on a movie ticket, $\frac{1}{3}$ of his money on a drink, and $\frac{1}{4}$ of his money on popcorn. What fraction of his money did he have left? Express the answer in simplest form.

Test A

Chapter 5 Multiplication of Fractions

Section A (2 points each)

Circle the correct option: **A**, **B**, **C**, or **D**.

1 $2 \times \dfrac{5}{6} =$ ⬚

   A $1\dfrac{2}{3}$ **B** $\dfrac{5}{12}$

   C $\dfrac{17}{5}$ **D** 2

2 $\dfrac{7}{9} \times 27 =$ ⬚

   A 3 **B** 21

   C $\dfrac{7}{27}$ **D** $\dfrac{9}{27}$

3 $1\frac{1}{3} \times 12 = $ ⬚

A 12 **B** 48

C 4 **D** 16

4 $\frac{3}{5} \times \frac{15}{16} = $ ⬚

A $\frac{5}{12}$ **B** $\frac{9}{16}$

C $\frac{5}{16}$ **D** $1\frac{1}{8}$

5 Which of the following is the reciprocal of the sum of $\frac{1}{4}$ and $\frac{1}{5}$?

A $\frac{1}{9}$ **B** $\frac{5}{4}$

C $\frac{20}{9}$ **D** $\frac{9}{20}$

Section B (2 points each)

Express all the answers in simplest form.

6 $\dfrac{1}{5} \times \dfrac{1}{6} =$

7 $\dfrac{5}{14} \times 2\dfrac{1}{10} =$

8 $\dfrac{1}{2} + 10 \times \dfrac{2}{5} =$

9 $2\dfrac{3}{4} \times 5\dfrac{1}{7} =$

10   Fill in the blank.

$1\dfrac{1}{9} \times \underline{\hspace{3em}} = 1$

11 Find the value.

$$25 + 18 \times \frac{1}{6} - 2\frac{1}{2} = \boxed{}$$

12 Find the value.

$$\frac{1}{3} \times \left(\frac{1}{4} + \frac{1}{4}\right) \times \frac{3}{5} = \boxed{}$$

13 There are 56 lemons and limes in a crate. $\frac{3}{8}$ of them are limes. How many limes are in the crate?

14 Dion ran $1\frac{1}{8}$ mile every day last week. How many miles did he run altogether last week?

Section C (4 points each)

Express all the answers in simplest form.

15 There was $\frac{7}{10}$ L of water in a bottle. $\frac{1}{2}$ of the water was poured out. How many liters of water are still in the bottle?

16 There were 188 people on a train. $\frac{1}{4}$ of the passengers got off the train at the first stop and 19 passengers got on at the same stop. How many passengers were on the train after the first stop?

17 A baker made 160 cupcakes. She sold $\frac{1}{2}$ of them in the morning and $\frac{4}{5}$ of the remainder in the afternoon. How many cupcakes did she sell in the morning and afternoon altogether?

40

Name: _______________________

Date: _______________________

Test B

Chapter 5 Multiplication of Fractions

Section A (2 points each)

Circle the correct option: **A**, **B**, **C**, or **D**.

1 $15 \times \frac{6}{75} =$ [______]

   A $6\frac{1}{5}$ **B** $15\frac{6}{75}$

   C $1\frac{1}{5}$ **D** $\frac{7}{25}$

2 $\frac{9}{7} \times 49 =$ [______]

   A 7   **B** $\frac{9}{49}$

   C 343   **D** 63

3 $\frac{7}{4} \times \frac{7}{4} =$

A $3\frac{1}{2}$

B $3\frac{1}{16}$

C 1

D $\frac{7}{16}$

4 $3\frac{3}{10} \times 2\frac{1}{11} =$

A $5\frac{3}{110}$

B $6\frac{4}{11}$

C $6\frac{9}{10}$

D $2\frac{3}{10}$

5 What is the sum of the reciprocals of $\frac{1}{4}$ and $\frac{4}{5}$?

A $5\frac{1}{4}$

B $\frac{20}{21}$

C $\frac{4}{9}$

D $4\frac{4}{5}$

Express all the answers in simplest form.

6 $\frac{1}{2} \times \frac{1}{3} \times \frac{1}{4} =$ ⬚

7 Find $\frac{2}{3}$ of $3\frac{3}{5}$.

8 $9 \times \left(\frac{1}{6} + \frac{1}{3}\right) - \frac{1}{6} \times 10 =$ ⬚

9 $\frac{2}{7} \times \frac{3}{10} \times 5 \times 7 =$ ⬚

10 Fill in the blank.

$$\underline{\hspace{2cm}} \times \frac{3}{14} \times \frac{2}{3} = 1$$

11 $6 \times \frac{3}{16} + \frac{1}{10} \times \frac{5}{8} =$ ☐

12 $36 \div (\frac{1}{12} \times 48) \times \frac{2}{3} =$ ☐

13 A bag of apples weighs $\frac{15}{16}$ lb. A bag of oranges weighs $1\frac{1}{2}$ times as much as the bag of apples. How many pounds does the bag of oranges weigh?

14 The price of a bicycle is \$150. The price of a bicycle helmet is $\frac{1}{6}$ the price of the bicycle. How much more does the bicycle cost than the helmet?

Section C (4 points each)

Express all the answers in simplest form.

15 A bottle can hold $1\frac{1}{5}$ cups of water. A kettle can hold $3\frac{1}{2}$ times as much water as the bottle. How many cups of water can the bottle and the kettle hold altogether?

16 There were 208 people on a train. $\frac{3}{8}$ of the passengers got off the train at the first stop. Of the passengers who got off, $\frac{5}{6}$ of them were adults and the rest were children. How many children got off the train at the first stop?

17 $\frac{2}{5}$ of the rodents sold by a pet store last week were hamsters. $\frac{4}{9}$ of the remaining rodents sold were gerbils and the rest of the rodents sold were mice. The store sold 54 hamsters. How many mice did the pet store sell last week?

Test A

Chapter 6 Division of Fractions

Section A (2 points each)

Circle the correct option: **A**, **B**, **C**, or **D**.

1 $\frac{1}{5} \div 8 =$ ⬚

    A $\frac{1}{8}$ **B** $\frac{8}{5}$

    C $\frac{1}{40}$ **D** 40

2 $\frac{2}{7} \div 2 =$ ⬚

    A $\frac{1}{7}$ **B** $\frac{1}{14}$

    C $\frac{4}{7}$ **D** $\frac{7}{4}$

3 $\dfrac{9}{5} \div 4 = \boxed{}$

A $\dfrac{36}{5}$ **B** $\dfrac{20}{9}$

C $\dfrac{9}{20}$ **D** $4\dfrac{5}{9}$

4 $12 \div \dfrac{1}{2} = \boxed{}$

A 6   **B** 24

C $\dfrac{1}{6}$ **D** $\dfrac{1}{24}$

5 6 friends shared $\dfrac{1}{2}$ a pizza equally. What fraction of the pizza did each friend get?

A $\dfrac{1}{6}$ **B** $\dfrac{1}{12}$

C $\dfrac{1}{2}$ **D** $\dfrac{1}{3}$

Section B (2 points each)

Express all the answers in simplest form.

6 $\frac{1}{3} \div 3 =$

7 $\frac{3}{5} \div 10 =$

8 $2 \div \frac{1}{6} =$

9 $5 \div \frac{2}{3} =$

10 $3 \div \frac{7}{2} =$

11 Find the value.

$$(\tfrac{1}{4} + \tfrac{1}{3}) \div 7 = \boxed{}$$

12 Find the value.

$$10 - 9 \div \tfrac{3}{2}$$

13 $6 \div \boxed{} = 18$

14 Emma cuts $\tfrac{9}{10}$ m of ribbon into 5 pieces of equal length. How long is each of the 5 pieces of ribbon in meters?

Section C (4 points each)

Express all the answers in simplest form.

15 14 identical bars of chocolate weigh $\frac{7}{10}$ kg. How many kilograms do 3 bars of chocolate weigh?

16 Maya put $\frac{2}{5}$ of a cake onto a plate and cut it into 6 pieces. Her sister ate 2 pieces of the cake from the plate. What fraction of the whole cake did Maya's sister eat?

 A juice shop made 24 L of lemonade and put $\frac{3}{4}$ of it in a juice dispenser. The rest of the lemonade was poured into bottles that hold $\frac{3}{8}$ L each. How many bottles of lemonade did the shop fill?

Test B

Chapter 6 Division of Fractions

Section A (2 points each)

Circle the correct option: **A**, **B**, **C**, or **D**.

1 $\frac{1}{9} \div 9 =$ [ ]

 A 9   **B** $\frac{1}{81}$

 C 81   **D** 1

2 $\frac{7}{8} \div 14 =$ [ ]

 A $\frac{1}{16}$ **B** $\frac{14}{8}$

 C $\frac{1}{4}$ **D** 4

3 $\dfrac{4}{3} \div 3 \times 2 =$ [   ]

    A 8   **B** $\dfrac{8}{9}$

    C $\dfrac{2}{9}$ **D** $\dfrac{4}{9}$

4 $40 \div \dfrac{1}{8} =$ [   ]

    A 5   **B** $\dfrac{1}{40}$

    C $\dfrac{1}{5}$ **D** 320

5 How many bags are needed to put 6 lb of candy into each bag so that there is $\dfrac{2}{3}$ lb of candy in each bag?

    A 4   **B** 12

    C 9   **D** $\dfrac{1}{9}$

Section B (2 points each)

Express all the answers in simplest form.

6 $\frac{1}{20} \div 5 = \boxed{}$

7 $\frac{11}{5} \div 8 = \boxed{}$

8 $6 \div \frac{1}{60} = \boxed{}$

9 $9 \div \frac{6}{11} = \boxed{}$

10 $20 \div \frac{16}{5} = \boxed{}$

11 $\left(\frac{2}{3} - \frac{1}{5}\right) \div 3 = \boxed{}$

12 $1\frac{3}{4} + \frac{8}{3} \div 2 = \boxed{}$

13 $\boxed{} \div \frac{1}{8} = 72$

14 Alex had a string $\frac{11}{12}$ ft long. He cut the string into equal pieces to form a square. What is the length of each of the pieces of string that he used to form the square?

Section C (4 points each)

Express all the answers in simplest form.

15 A bag of pecans weighs $\frac{6}{10}$ kg. A box containing some bags of pecans weighs $9\frac{1}{4}$ kg. The box itself weighs $\frac{1}{4}$ kg. How many bags of pecans are in the box?

16 Cody gave $\frac{1}{4}$ of a cake to his neighbor. He cut the rest of the cake into 9 equal pieces and put 5 pieces in the refrigerator. What fraction of the whole cake did he put in the refrigerator?

17 $\frac{3}{4}$ lb of corn nuts cost \$6. Sara bought $3\frac{1}{2}$ lb of corn nuts and Pablo bought $1\frac{5}{8}$ lb of corn nuts. How much more money did Sara spend on corn nuts than Pablo?

Name: _______________________________

Date: _______________________________

40

Test A

Chapter 7 Measurement

Section A (2 points each)
Circle the correct option: **A**, **B**, **C**, or **D**.

1 $\frac{3}{5}$ min = ⬚ s

A 36 **B** 35

C 12 **D** 180

2 $4\frac{2}{3}$ ft = ⬚ in

A 8 **B** 48

C 56 **D** 42

3 Han used $\frac{2}{5}$ kg of chocolate chips to make cookies. How many grams of chocolate chips did he use?

A 250 g

B 200 g

C 400 g

D 25 g

4 What is the area of this figure?

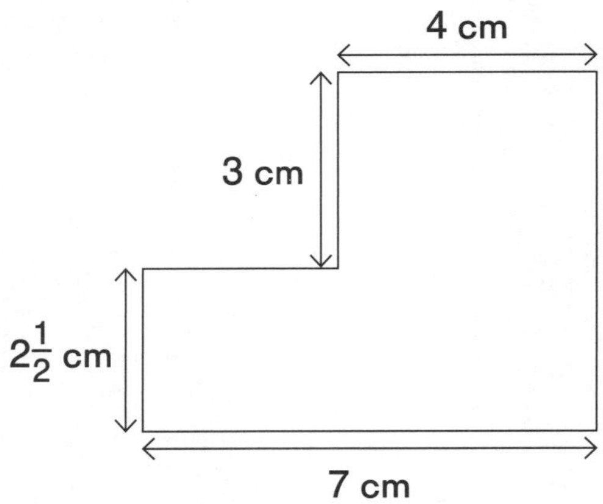

A $17\frac{1}{2}$ cm²

B $36\frac{1}{2}$ cm²

C $5\frac{1}{2}$ cm²

D $29\frac{1}{2}$ cm²

5 What is the area of Triangle ACD?

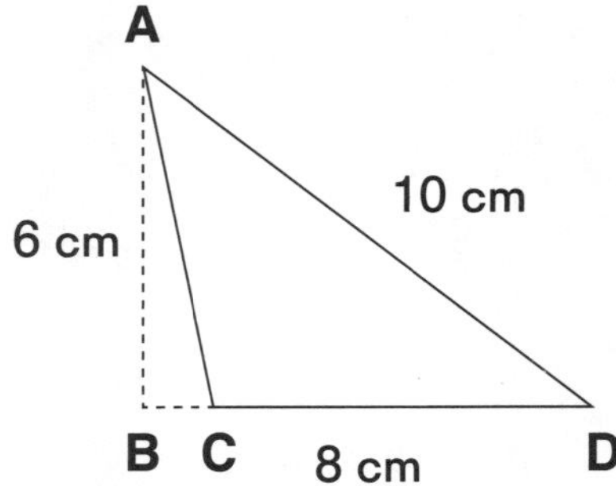

A 48 cm²

B 24 cm²

C 40 cm²

D 30 cm²

Section B (2 points each)

Express all the answers in simplest form.

6 $5\frac{1}{4}$ lb = _________ lb _________ oz

7   Express $7\frac{1}{5}$ km in meters.

8   Find the area of this square.

9 Find the area of the shaded triangle in the figure below.

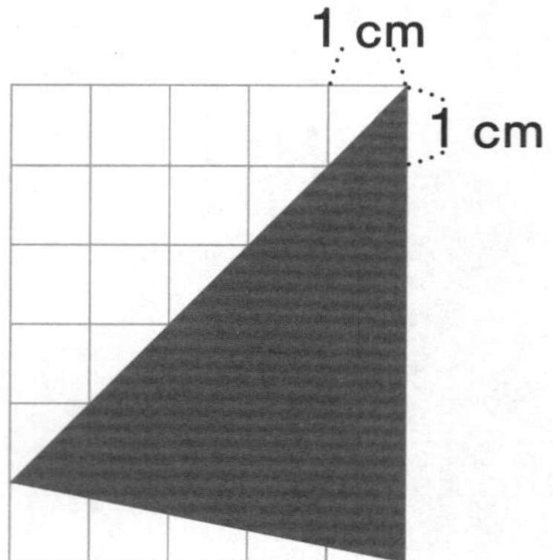

10 Identify the base of Triangle QRS for the given height.

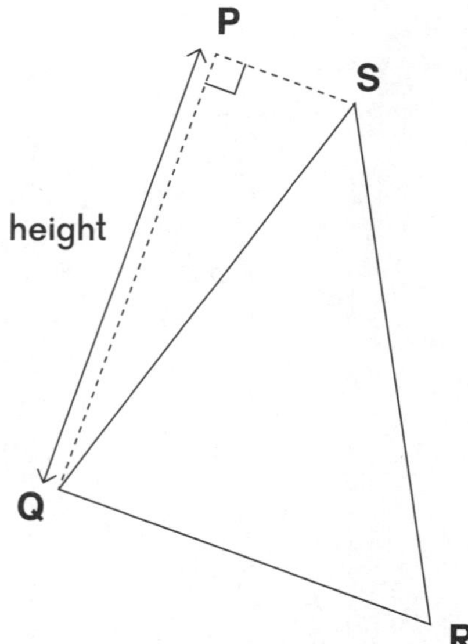

11 If MN is the base of Triangle MNO, what is the height?

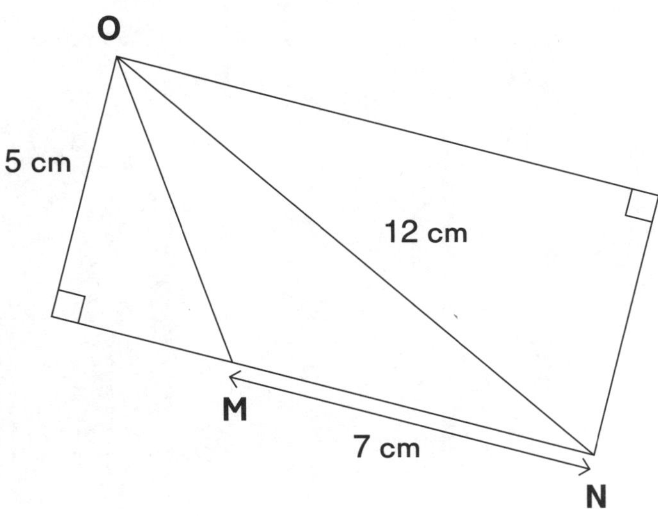

12 Find the area of Triangle DEF.

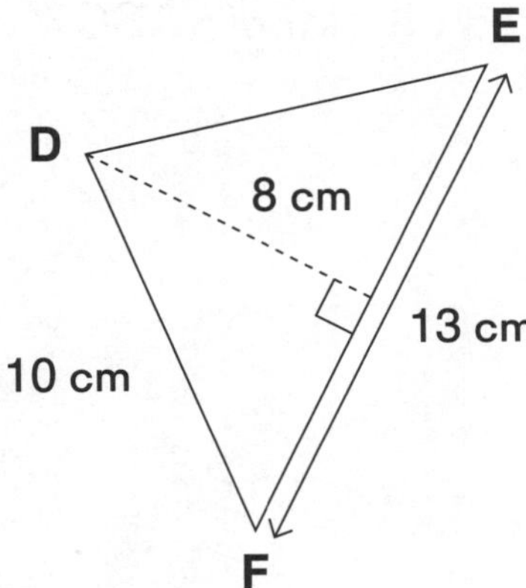

13 Find the area of Triangle XYZ.

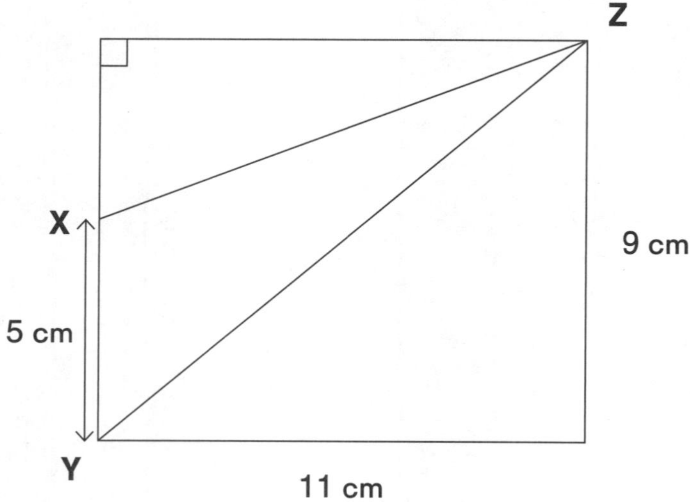

14 Triangles ABC and DBC share the same base, BC. The height of Triangle ABC is twice that of Triangle DBC. What is the area of Triangle ABC?

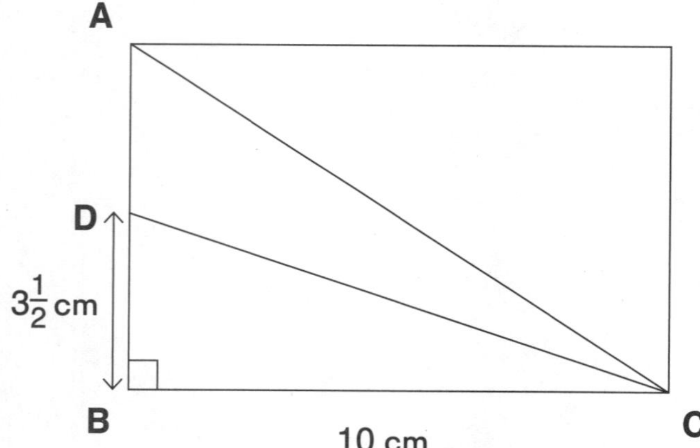

Section C (4 points each)

Express all answers in simplest form.

15 Emma left her house at 2:00 p.m. to go to the library. It took her $\frac{3}{4}$ h to get to the library. She spent $1\frac{1}{12}$ h at the library. What time did she leave the library?

16 A rectangular pond is surrounded by a lawn. Find the area of the lawn.

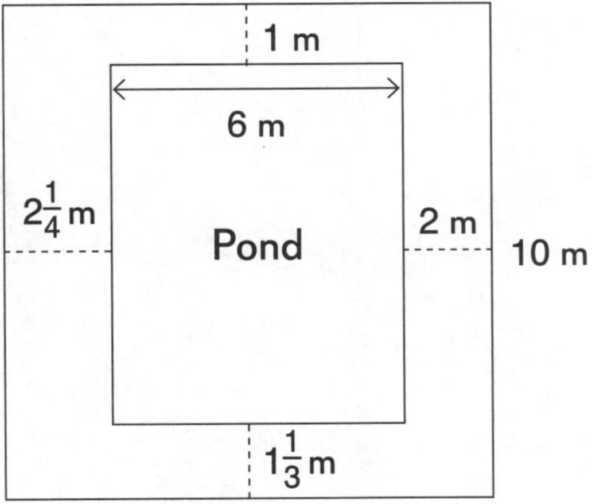

 Find the area of the shaded part.

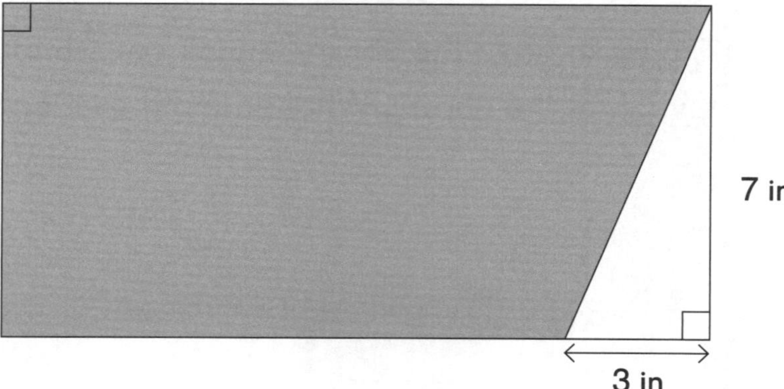

Test B

Chapter 7 Measurement

Section A (2 points each)
Circle the correct option: **A**, **B**, **C**, or **D**.

1 $\frac{7}{8}$ lb = [　] oz

A 2

B 9

C 78

D 14

2 $6\frac{1}{4}$ L = [　] mL

A 6,140

B 6,250

C 24

D 6,025

3 One bag of candy weighs $\frac{3}{4}$ kg. What is the weight of two bags of candy in grams?

A 750 g

B 12 g

C 1,500 g

D 340 g

4 Which triangle has an area three times the area of Triangle ABC?

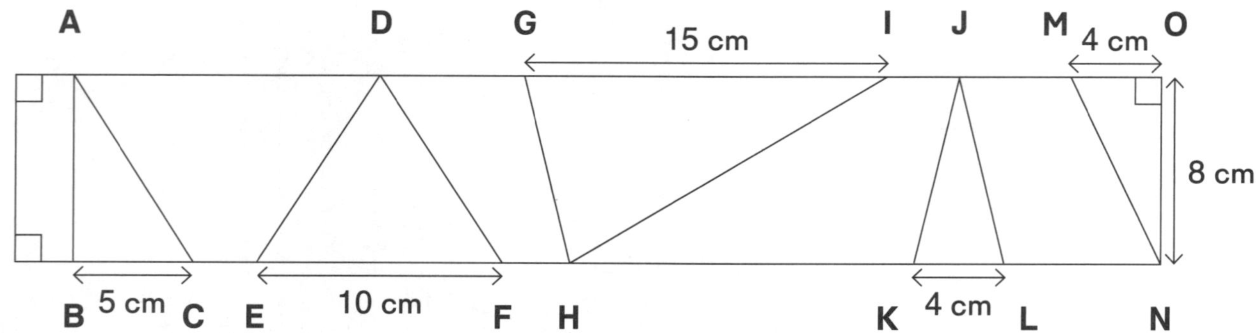

A MNO

B GHI

C DEF

D JKL

5 What is the area of Triangle PQR?

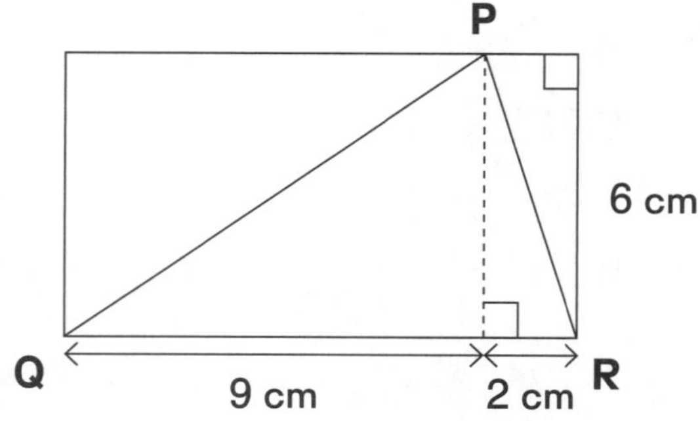

A 33 cm²

B 18 cm²

C 66 cm²

D 6 cm²

Express all the answers in simplest form.

6 $10\frac{1}{20}$ m = _______ m _______ cm

7   Express $3\frac{3}{4}$ gal in quarts.

8   Find the area of this rectangle.

$1\frac{1}{2}$ in

$4\frac{2}{9}$ in

9 Identify the height of Triangle EFG for the base FG.

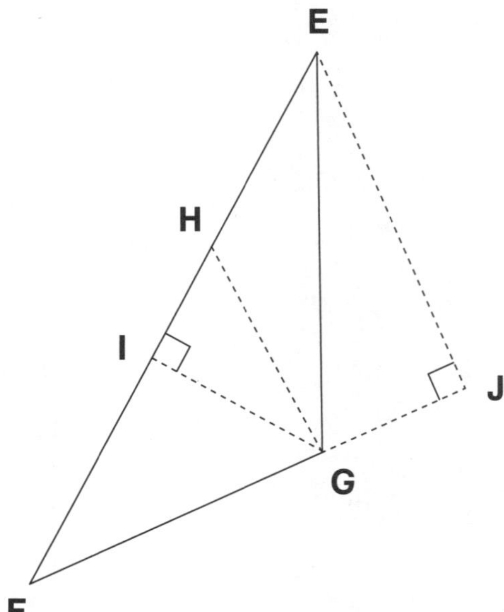

10 Find the area of the shaded part.

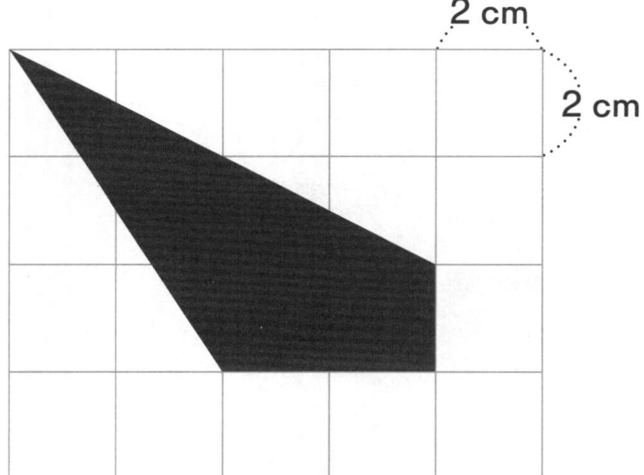

11 Find the area of Triangle XYZ.

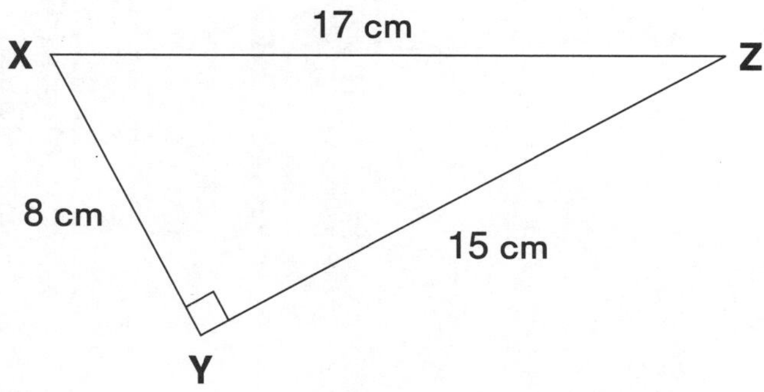

12 Find the area of Triangle STU.

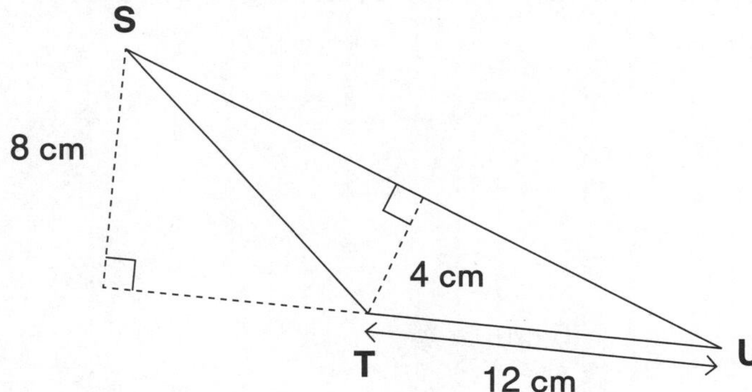

13 Find the area of the shaded part of the following figure.

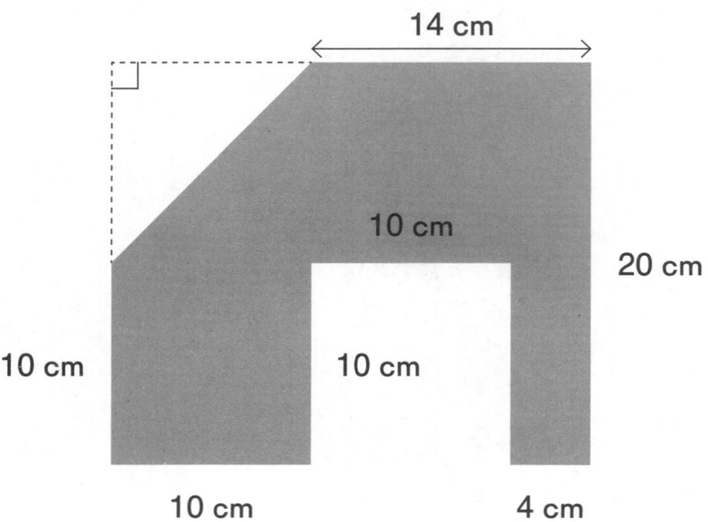

14 Figure EFGH has an area twice that of Triangle EFG. What is the area of Figure EFGH?

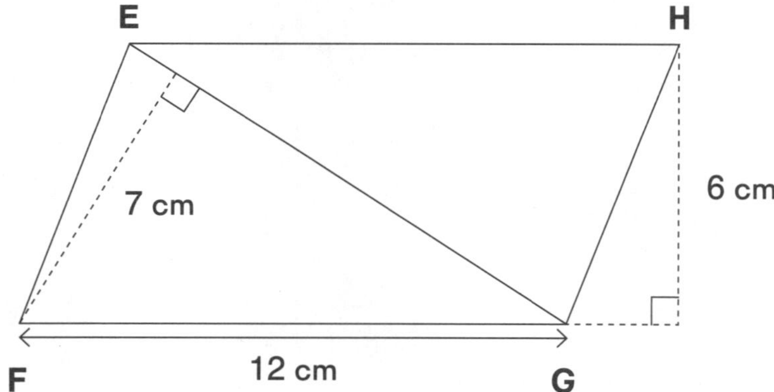

15 Jacob boarded a train $\frac{1}{3}$ hours before it departed from the station. He arrived at his destination $2\frac{5}{12}$ hours after the train left the station. It was 10:30 a.m. when he boarded the train. What time did he arrive at his destination?

16 Find the area of the shaded part.

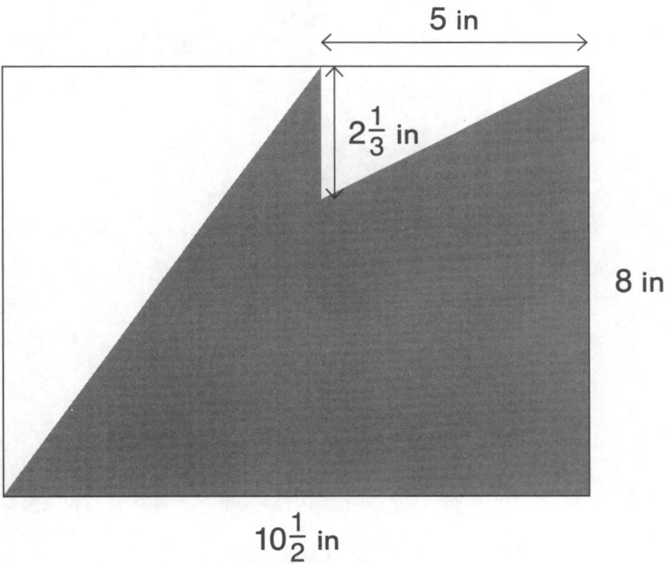

17 A triangle with a base of 24 mm and a height of 2 cm has the same area as a rectangle. What is the perimeter of the rectangle in millimeters if the length of one of its sides is $2\frac{1}{2}$ cm?

25 min **Score**

30

Test A

Chapter 8 Volume of Solid Figures

Section A (2 points each)
Circle the correct option: **A**, **B**, **C**, or **D**.

1 How many cubes were added to the solid on the left to create the solid on the right?

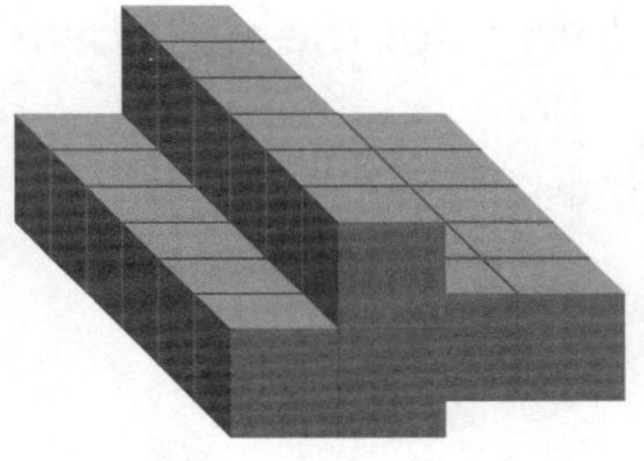 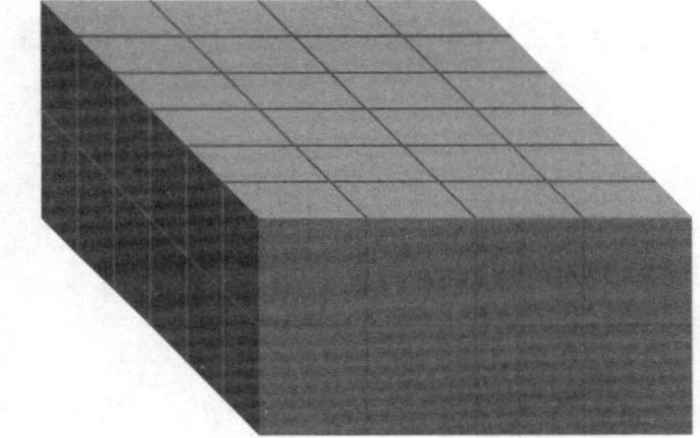

A 16 **B** 20

C 18 **D** 48

2 A rectangular box measures 10 cm by 10 cm by 10 cm. How many 1-cm cubes are needed to fill this box?

A 1,000 **B** 100

C 30 **D** 3,000

3 What is the volume of this cuboid?

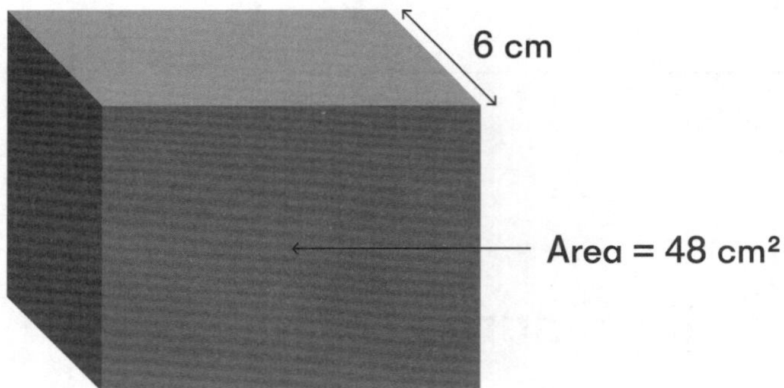

A 54 cm³

B 288 cm³

C 36 cm³

D 42 cm³

4 The volume of this cuboid is 660 cm³. What is the length of the unknown edge?

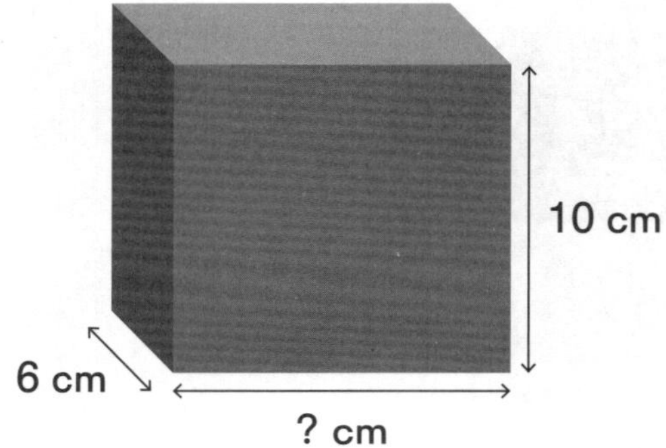

A 66 cm

B 110 cm

C 165 cm

D 11 cm

5 2 L 85 mL = ☐ cm³

A 285

B 2,850

C 2,085

D 105

6 Find the volume of this cuboid.

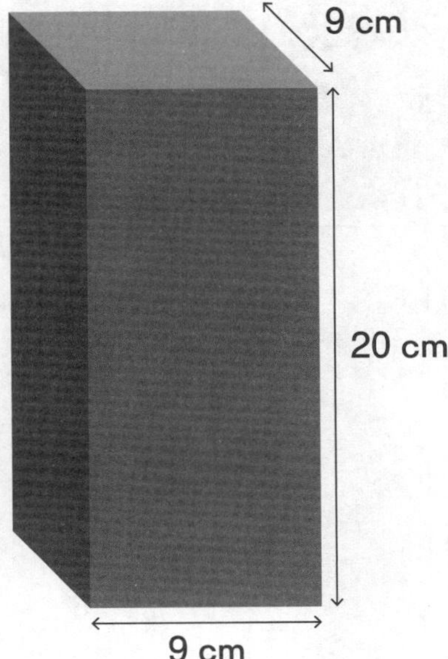

7 The volume of the following cuboid is 336 cm³. What is the length of the unknown edge?

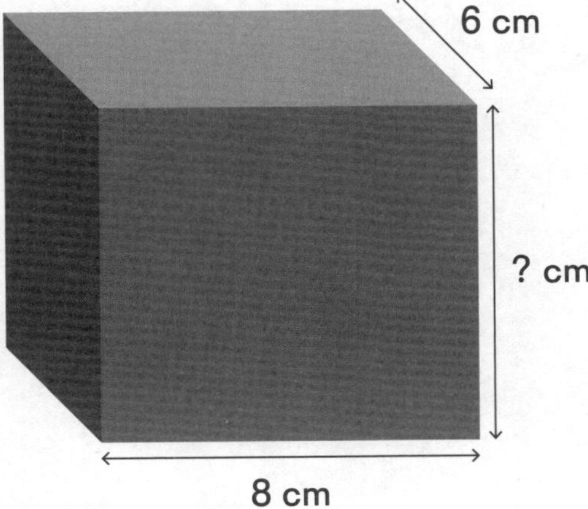

8 The volume of this cuboid is 1,250 in³. Find the length of the unknown edge.

9 Fill in the blanks.

(a) 575 mL = _________ cm³

(b) 1,950 cm³ = _________ L _________ mL

 Find the volume of this solid figure.

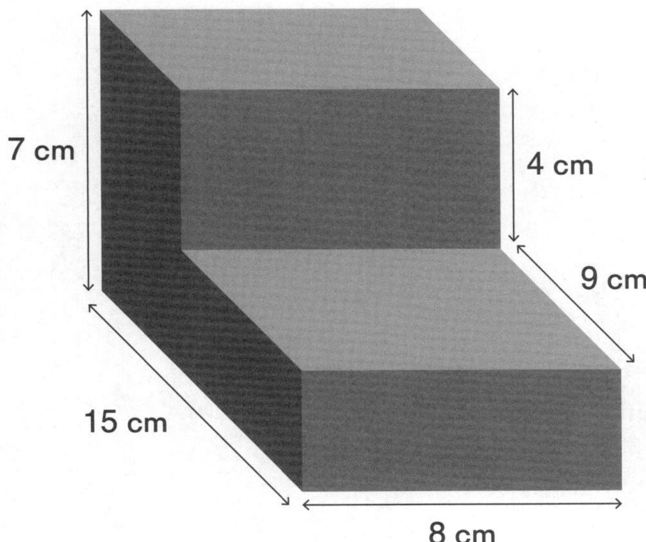

11 The water level in a rectangular tank with a brick in it is 19 cm. The tank's base is 20 cm by 30 cm. When the brick was removed from the tank, the water level decreased to 17 cm. What is the volume of the brick?

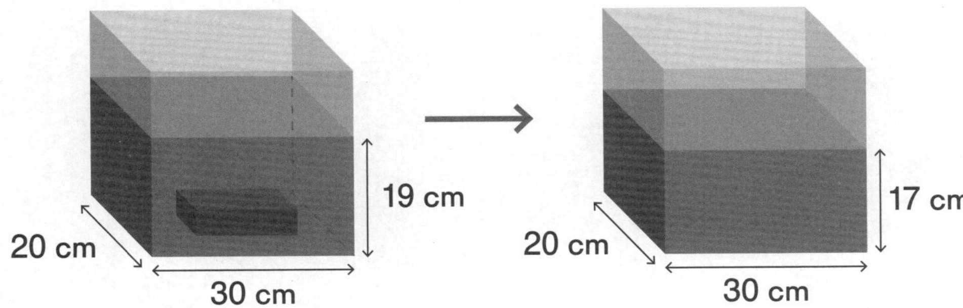

12 Find the volume of this solid figure.

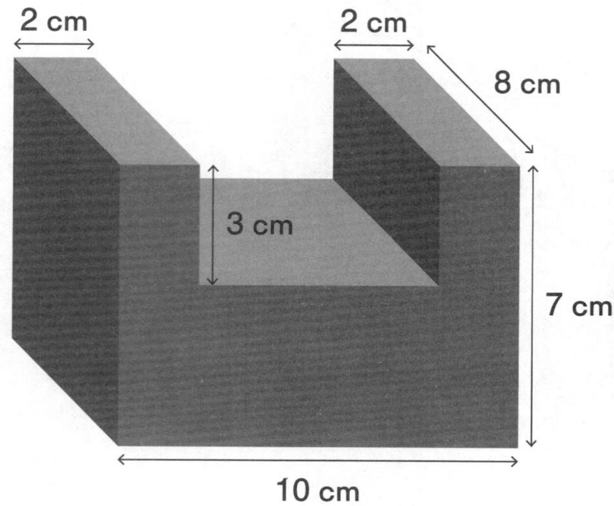

13 A rectangular tank with a base 10 in by 8 in contains water to a height of 6 in. A rock with a volume of 240 in^3 is placed in the tank. What is the new height of the water in the tank?

25 min   **Score**

30

Test B

Chapter 8 Volume of Solid Figures

Section A (2 points each)
Circle the correct option: **A, B, C,** or **D.**

1 How many cubes were removed from the solid on the left to create the solid on the right?

 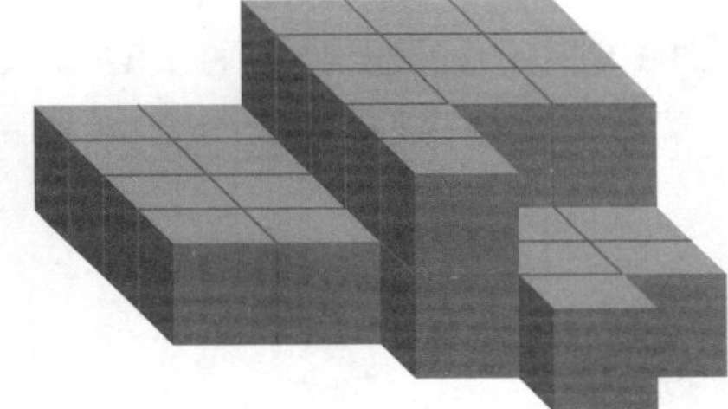

A 8   **B** 11

C 12   **D** 13

2 A rectangular box measures 12 cm by 10 cm by 9 cm. How many 1-cm cubes are needed to fill half of this box?

A 120   **B** 31

C 540   **D** 1,080

 The volume of this cube is 512 cm³. What is the length of one of the edges?

A 128 cm

B 8 cm

C 16 cm

D 3 cm

4 What is the volume of this solid figure?

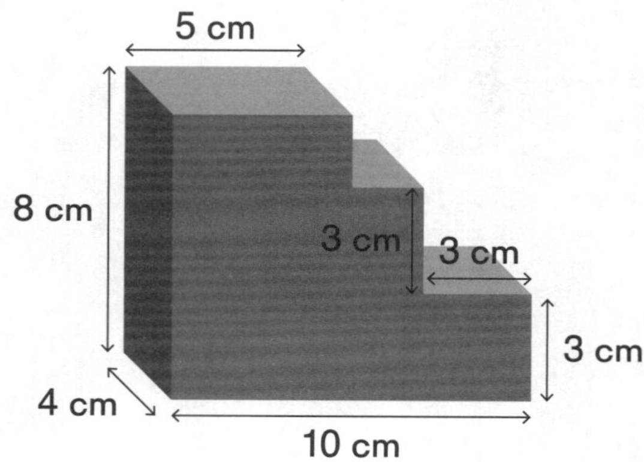

A 244 cm³

B 200 cm³

C 184 cm³

D 196 cm³

5 $3\frac{1}{2}$ L = ☐ cm³

A 350

B 3,500

C 35

D 7

6 Find the volume of this cuboid.

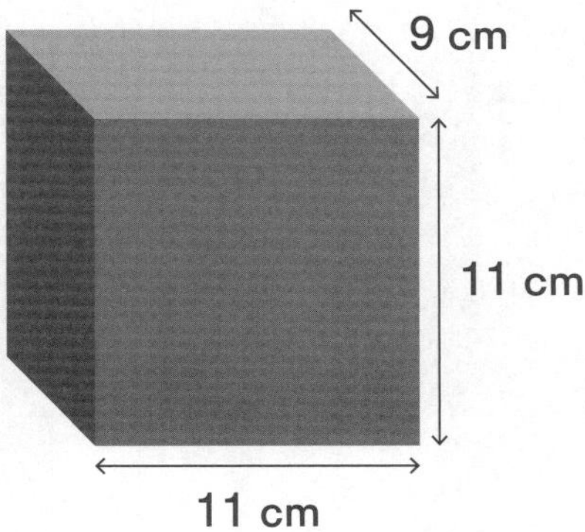

7 The volume of the following cuboid is 6,500 cm³. What is the length of the unknown edge?

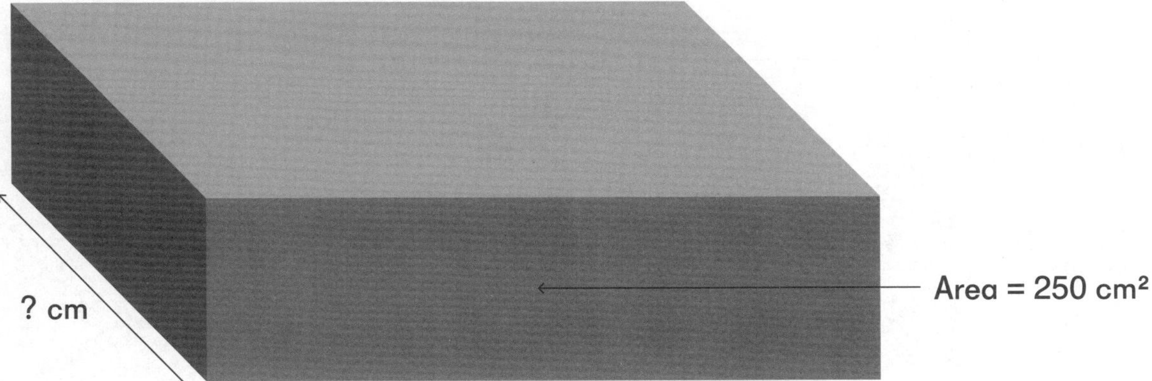

8 The volume of a rectangular storage bin is 1,280 cm³. The area of its base is 80 cm². What is the height of this container?

9 Fill in the blanks.

(a) 3 L 5 mL = _________ cm³

(b) 9,050 cm³ = _________ L _________ mL

10 A rectangular tank with a length of 25 cm and a width of 8 cm is $\frac{2}{3}$ filled with water to a height of 10 cm. What is the capacity of this tank in liters?

11 A rectangular tank with a base of 32 cm by 20 cm and a height of 14 cm was filled with water. Water was drained from the tank until the water level fell to 8 cm. How many liters and milliliters of water was drained from the tank?

12 Find the volume of this solid figure.

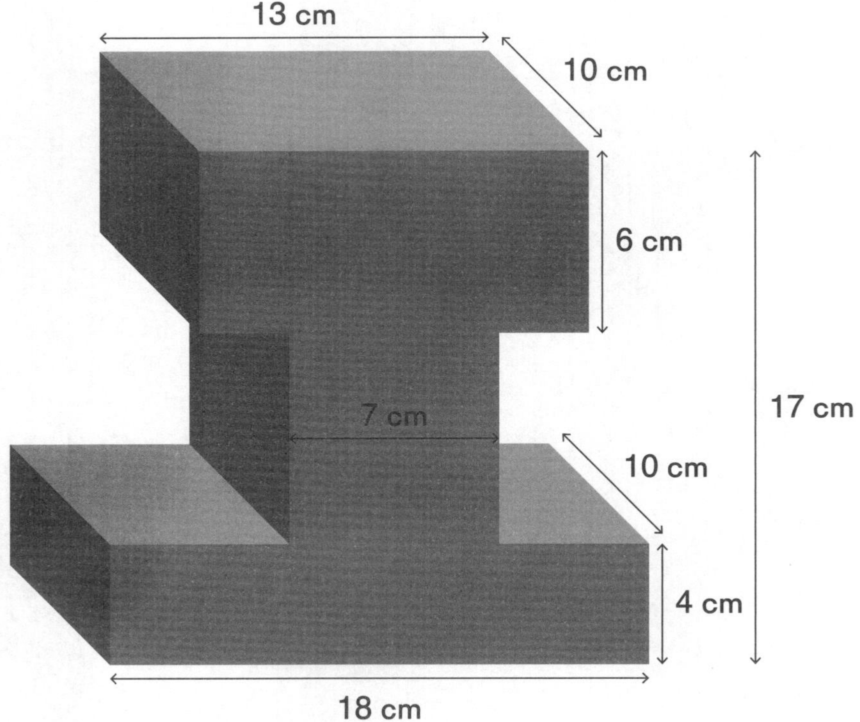

13 A rectangular tank has a base 10 in by 9 in and a height of 8 in. It is half filled with water. 60 identical metal balls, each with a volume of 3 in³ were dropped into the tank. What is the new height of the water level in the tank?

Name: ______________________________________

Date: ______________________________________

Score: 60

Test A

Continual Assessment 2

Section A (2 points each)
Circle the correct option: **A**, **B**, **C**, or **D**.

1 One hundred million less than 538,201,760 is ______________.

  A 538,101,760   **B** 438,201,760

  C 538,201,660   **D** 638,201,760

2 $70,000 \times 500 =$ ☐

  A 35,000   **B** 350,000

  C 35,000,000   **D** 3,500,000

3 $(25 + 10) \times 2 - 1 =$ ☐

  A 35   **B** 44

  C 69   **D** 45

4 $18 - 6 \div 2 \times 5 =$

A 3

B 30

C 75

D 10

5 $8{,}723 \times 82 =$

A 697,840

B 696,000

C 715,040

D 715,286

6 $\frac{3}{7} + 2\frac{1}{2} =$

A $2\frac{4}{9}$

B $\frac{13}{14}$

C $2\frac{13}{14}$

D $2\frac{4}{7}$

7 $60 - 20 \times \frac{1}{4} =$

A 55

B 65

C 40

D 10

8 $\frac{5}{8} \times \frac{16}{7} =$ ☐

A 10

B $1\frac{3}{7}$

C $\frac{2}{7}$

D $5\frac{2}{7}$

9 $4\frac{3}{5}$ km = ☐ m

A 460

B 4,350

C 4,300

D 4,600

10 What is the area of Triangle MNO?

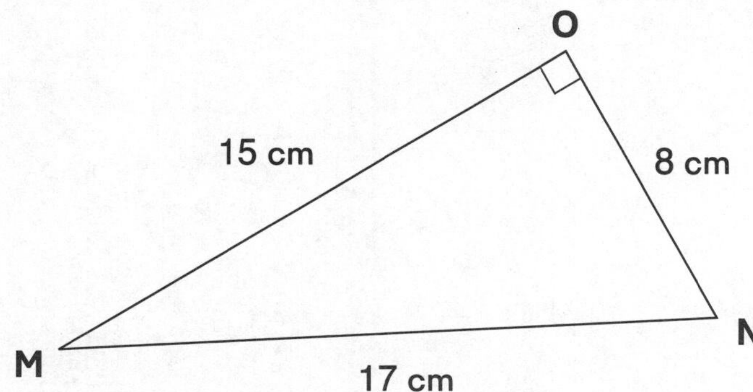

A 68 cm²

B 120 cm²

C 60 cm²

D 30 cm²

Section B (2 points each)

Express all the answers in simplest form.

11 $630,000 \div 1,000 = $ ☐

12 $270,505,108 = 70,000,000 + 200,000,000 + $ ☐ $+ 108$

13 $82 \times 38 = 82 \times 40 - 82 \times$ ☐

14 Divide 784 by 26.

15 $10 \div \frac{2}{5} =$ [   ]

16 $7\frac{2}{3} - 4\frac{1}{5} =$ [   ]

17 55 concert tickets cost \$1,045. How much does 1 concert ticket cost?

18 Find the area of Triangle XYZ.

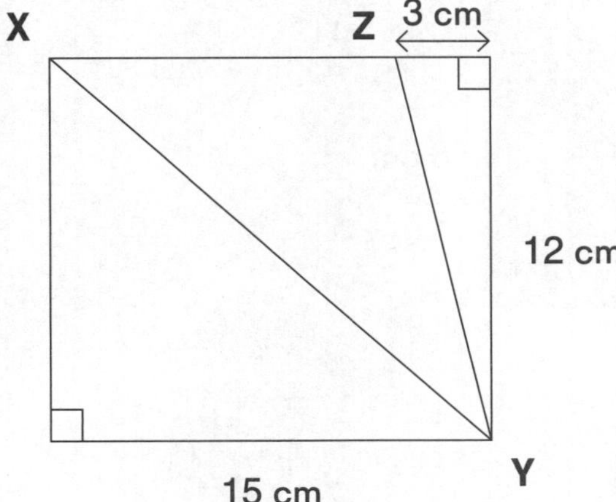

19 Find the area of this figure.

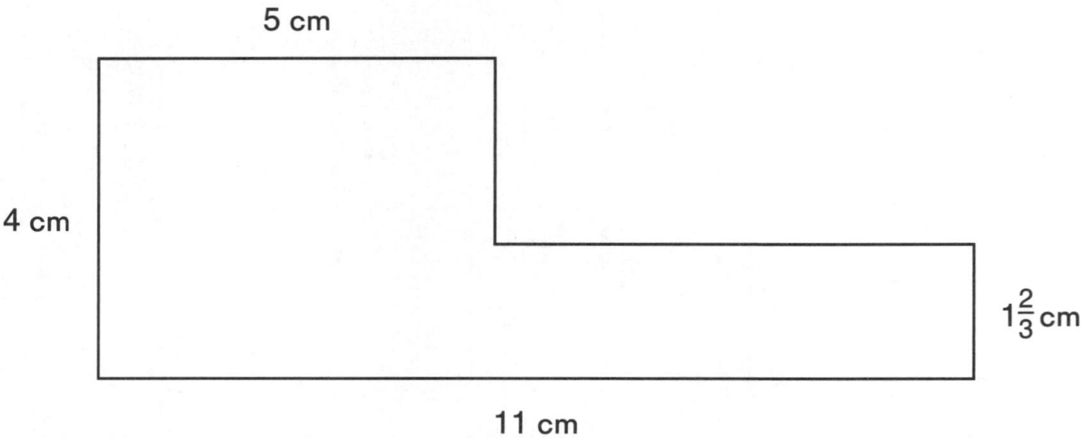

20 A rectangular tank, 25 cm by 10 cm by 8 cm, is half filled with water. How many more liters of water need are needed to fill the tank completely?

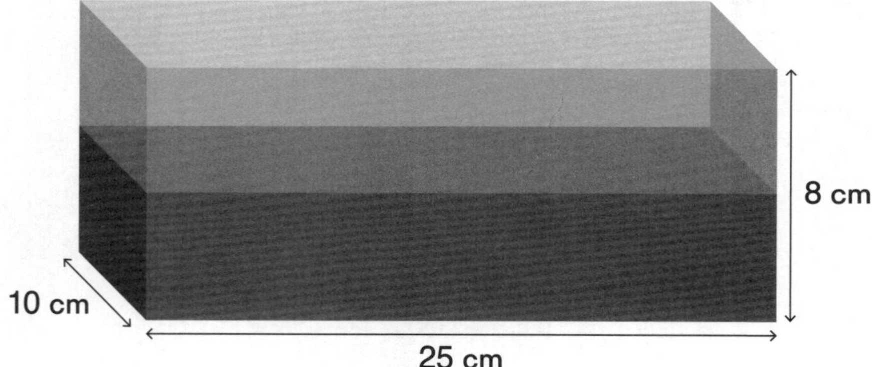

Section C (4 points each)

Express all the answers in simplest form.

21 Alex and Emma had the same number of stickers at first. Alex bought 15 more stickers and Emma bought 3 times as many stickers as Alex bought. They now have a total of 78 stickers. How many stickers did each of them have to begin with?

22 A juice vendor made 12 L of carrot juice and filled some bottles each with $\frac{2}{5}$ L of the juice. He sold all the bottled carrot juice for $6 each. How much did he receive from selling the carrot juice?

23 A rectangular window measures $10\frac{1}{3}$ ft by 8 ft. What is the area of this window in square inches?

24 $\frac{3}{20}$ of the people at a basketball game are children. There are 1,680 more adults than children at the game. How many children are at the game?

25 Find the volume of this solid.

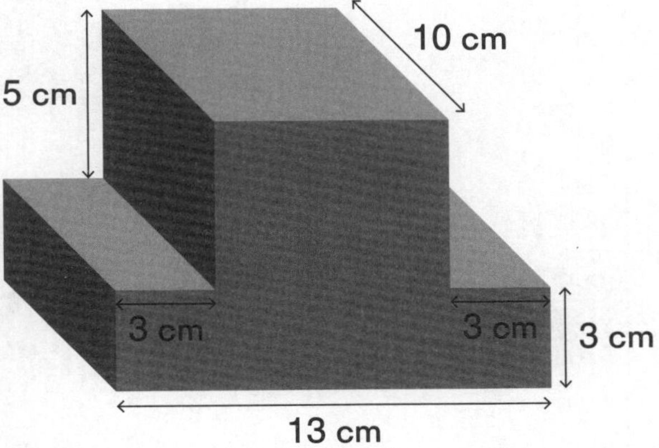

BLANK

Test B

Continual Assessment 2

Section A (2 points each)

Circle the correct option: **A**, **B**, **C**, or **D**.

1 88,000,000 is _______________ times as great as 880.

A 10,000

B 10,000,000

C 100,000

D 1,000,000

2 1,705,000 ÷ 100 = ☐

A 17,050

B 1,705

C 170,500

D 170,500,000

3 $100 - 20 \div 4 + 5 \times 2 - 5 = $ ☐

A 25

B 195

C 100

D 90

4 $(2 \times 20 - 10) \div 5 + 5 =$ ⬚

   A 9   **B** 2

   C 3   **D** 11

5 A factory makes 7,500 lb of dog treats per day. How many pounds of dog treats does the factory make in 31 days?

   A 250 lb   **B** 232,500 lb

   C 225,000 lb   **D** 240,000 lb

6 $2\frac{2}{3} - \left(\frac{1}{3} + \frac{1}{5}\right) =$ ⬚

   A $2\frac{2}{15}$ **B** $2\frac{8}{15}$

   C $\frac{8}{15}$ **D** $1\frac{7}{15}$

7 $\frac{6}{5} \times \frac{6}{7} =$ ⬚

   A $1\frac{2}{5}$ **B** $1\frac{1}{35}$

   C $\frac{5}{7}$ **D** $\frac{35}{36}$

8 $(4 - 3\frac{2}{7}) \div 3 =$ ☐

A $3\frac{6}{7}$ **B** $\frac{2}{21}$

C $\frac{5}{21}$ **D** $2\frac{1}{7}$

9 What is the reciprocal of $3\frac{2}{7}$?

A $\frac{7}{23}$ **B** $\frac{23}{7}$

C $\frac{7}{2}$ **D** $3\frac{7}{2}$

10 What is the height of Triangle DEF given that the base is DE?

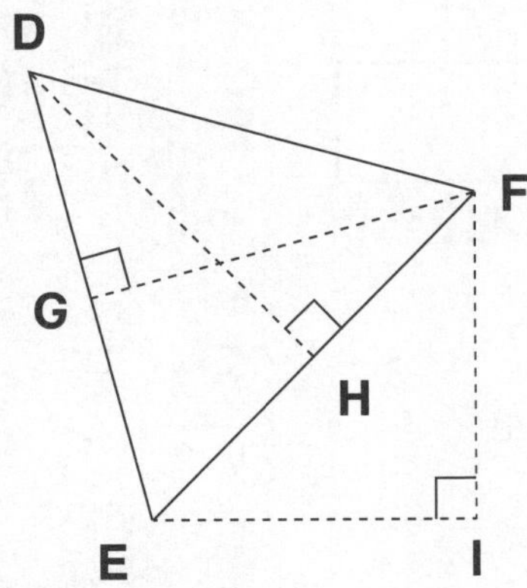

A EI   **B** FI

C GF   **D** DH

Express all the answers in simplest form.

11 738,200,000 has _______________ ten thousands.

12 Write >, <, or = in the ◯.

20,600,000 + 7,000 ◯ 670,000 + 700 + 20,000,000

13 345 × 9,003 = ☐ × 9,000 + 345 × ☐

14 17 × 115 × 6 = ☐

15 $6 \div \dfrac{8}{3} =$

16 $\dfrac{2}{5} \div 12 + 1 \div \dfrac{1}{15} =$

17 A bakery made 1,450 cookies and packed 16 cookies in each box. How many cookies were not packed in boxes?

18 There were $2\dfrac{3}{4}$ lb of pecans in a bag. Sara used $\dfrac{1}{2}$ of them to make cookies. How many ounces of pecans did she use?

19 Find the area of Triangle PQR.

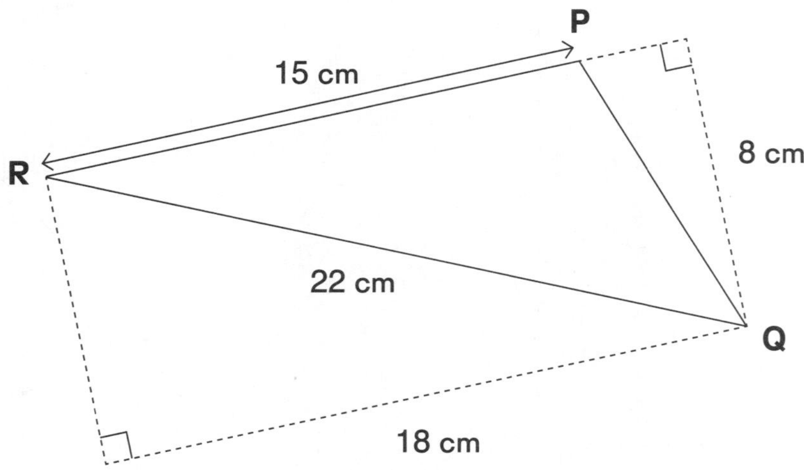

20 ABCD is a square. Find the area of the shaded part.

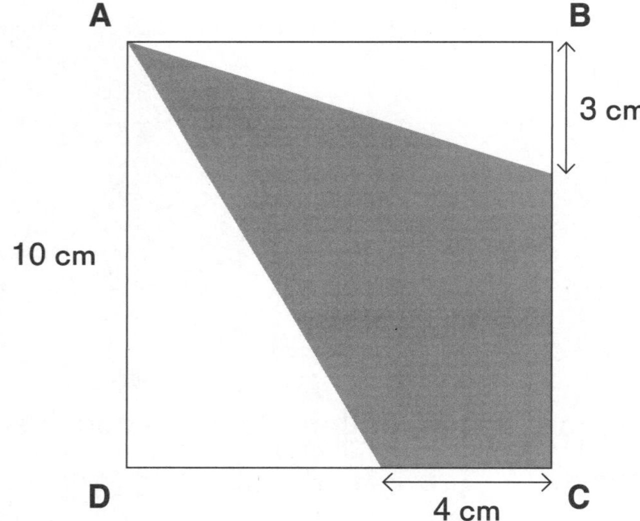

Section C (4 points each)

Express all the answers in simplest form.

21 A baker made 192 blueberry muffins and 64 lemon poppy muffins yesterday. The muffins were packed 4 in each box and each box was sold for $6. At the end of the day she had 28 muffins left. How much money did the baker receive selling muffins yesterday?

22 Dion saved $18 last month. He spent $\frac{1}{3}$ of it on an ice-cream cone. He spent the rest of his money on 6 sport cards that are equally priced. What fraction of his saving was the cost of each sport card?

23 $\frac{1}{3}$ lb of almonds costs \$3 and $1\frac{3}{4}$ lb of cashews costs \$21. How much more does one pound of cashews cost than one pound of almonds?

24 A cuboid has four square corners of the same size cut from it to form the solid below. Find the volume of this solid.

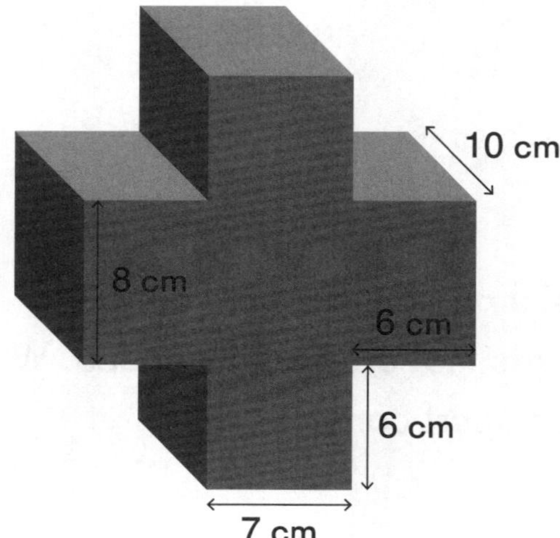

 The area of the base of a rectangular container is 420 cm². It was filled with water to a height of 9 cm. Some identical cuboids with sides 7 cm, 5 cm, and 2 cm were added and the height of the water rose to 11 cm. How many cuboids were added to the tank?

BLANK

Answer Key

Detailed solutions given are suggestions, and do not include all possible methods of arriving at the correct answer. Accept all reasonable solutions by students.

Chapter 1 Whole Numbers

1 B

2 C

3 D

4 B

5 A

6 Ten million, nine hundred five thousand, one hundred twenty-two

7 1,034,578

8 100,000

9 583,920,119

10 B, A, D, C

11 <

12 6,400,000

13 10,000

14 30,000

15 4,300

16 140,000

17 3,000

18 70

19 $280,000

20 200 bags

Chapter 1 Whole Numbers

1 A

2 C

3 B

4 D

5 B

6 583,010,011

7 9,652,001

8 206,000

9 8,427

10 50 hundred thousands = 5,000,000

five hundred millions = 500,000,000

5 ten millions = 50,000,000

five hundred millions, 5 ten millions, 50 hundred thousands

11 15,000,000 + 90,000 + 333 = 15,090,333

15,090,332

12 5,300,000

13 40,000,000

14 1,000,000

15 90,000,000

16 4,000,000

17 71,800

18 =

19 600,000 ÷ 5,000 = 120

120 envelopes

20 4,000 × 21 = 84,000

$84,000

Chapter 2 Writing and Evaluating Expressions

1 B

2 A

3 C

4 B

5 D

6 $36 \div 18 = 2$
2

7 $90 - 80 + 10 = 10 + 10 = 20$
20

8 $8 \times (20 - 10) = 8 \times 10 = 80$
80

9 $2 \times 12 \times 5 = 24 \times 5 = 120$
120

10 9; 12

11 29

12 2

13 >

14 $10 \times (3 + 7) = 100$

15 $3 \times 2 = 6$
$7 + 6 + 12 = 25$
$55 - 25 = 30$
$30

16 $100 \times 12 = 1{,}200$
$75 \times 6 = 450$
$1{,}200 + 450 - 15 = 1{,}650 - 15$
$= 1{,}635$
1,635 eggs

17 Regular / Child / Senior

230

10

10 units $\longrightarrow$ 230 + 10
1 unit $\longrightarrow$ 24
2 units $\longrightarrow$ 48
$48

Chapter 2 Writing and Evaluating Expressions

1 C

2 D

3 C

4 B

5 D

6 $180,000 \div 30,000 = 6$
6

7 $3 + 3 + 3 - 9 = 0$
0

8 $24 - 24 \div 3 \div 2 = 24 - 8 \div 2 =$
$24 - 4 = 20$
20

9 $4 \times 3 \times (6 \div 2 + 7) + 10 \div 2$
$= 4 \times 3 \times (3 + 7) + 5$
$= 4 \times 3 \times 10 + 5$
$= 12 \times 10 + 5$
$= 120 + 5 = 125$
125

10 100; 900

11 6,500

12 124; 3

13 $=$

14 $(6 + 4) \times (10 \div 2) = 50$

15 $(10 \times 2) + (10 \times 4) + 12$
$= 20 + 40 + 12 = 72$
$72 \div 10 = 7.20$
$\$7.20$

Chapter 2 Writing and Evaluating Expressions

16

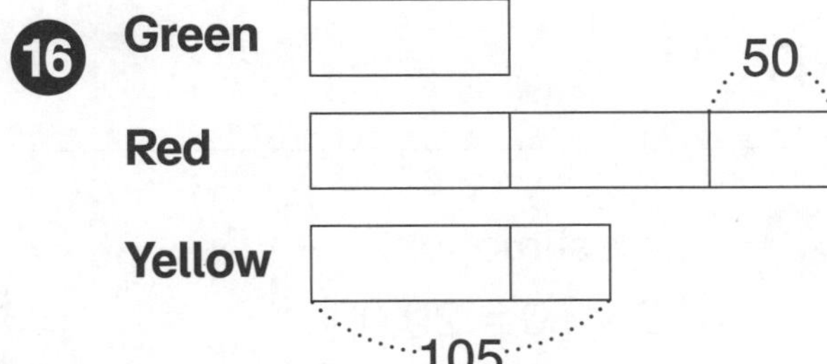

There are 2 units plus 50 red balls. The number of yellow balls is half of the number of red balls, so there is 1 unit and 25 yellow balls.

1 unit $\longrightarrow$ 105 − 25 = 80
4 units $\longrightarrow$ 80 × 4 = 320
320 + 50 + 25 = 395
395 balls

17 Before:

At first, Sofia has 2 units and Mei has 1 unit. Since Mei uses $\frac{1}{3}$ of her stickers, divide each unit into 3. Sofia now has 6 units and Mei has 3 units. Mei uses 1 of her units and has 2 units left. Sofia gives her 10 stickers, so Mei now has 2 units + 10. Since Sofia still has twice as many stickers as Mei, Sofia now has 4 units + 20. 4 units + 20 + the 10 she gave away must equal the 6 units she started with. Therefore, 2 units is 20 + 10, or 30 stickers.

2 units $\longrightarrow$ 30
1 unit $\longrightarrow$ 15
9 units $\longrightarrow$ 9 × 15 = 135
135 stickers

Chapter 3 Multiplication and Division

1 B

2 A

3 C

4 D

5 B

6 11,495

7 162,792

8 10,500

9 3 R 10

10 4 R 11

11 28 R 19

12 96

13 Use estimation.
$400 \times 50 = 20,000$
$15,000 + 3,000 = 18,000$
$>$

14 $2,645 \div 60$ is 44 R 5
44 h 5 min

15 $180 \times 15 = 2,700$
2,700 kg

16 $956 \div 24$ is 39 R 20
39 counters, 20 left over

17 $3,990 \div 95 = 42$
$42 \div 2 = 21$
$38 \times 21 = 798$
$798

Chapter 3 Multiplication and Division

1 C

2 D

3 C

4 B

5 C

6 30,305

7 387,504

8 50,061

9 3 R 17

10 1,201

11 4 R 71

12 23 R 1

13 Use estimation.
$50 \times 30 = 1,500$
$70,000 \div 70 = 1,000$
>

14 $4,810 \div 12$ is 400 R 10
400 ft 10 in

15 $685 + 208 = 893$
$893 \times 14 = 12,502$
12,502 lb

16 $500 \times 16 = 8,000$
$8,000 \div 21$ is 380 R 20
20 oz = 1 lb 4 oz
380 containers
1 lb 4 oz left unpacked

17 $15 + 75 = 90$
$90 \div 2 = 45$
$45 \times 23 + 15 = 1,050$
1,050 counters

Chapter 4 Addition and Subtraction of Fractions

1 C

2 A

3 D

4 C

5 D

6 $32\frac{1}{2}$

7 $3\frac{5}{6}$

8 $6\frac{1}{15}$

9 $9\frac{1}{6}$

10 $\frac{1}{12}$

11 $1\frac{13}{14}$

12 $2\frac{1}{5}$

13 $1\frac{5}{6}$

14 $3\frac{5}{6}$

15 $(5 + 9) \div 12 = \frac{14}{12} = 1\frac{2}{12} = 1\frac{1}{6}$

$1\frac{1}{6}$ L

16 $4\frac{1}{4} + 1\frac{1}{8} = 4\frac{2}{8} + 1\frac{1}{8} = 5\frac{3}{8}$

$4\frac{1}{4} + 5\frac{3}{8} = 4\frac{2}{8} + 5\frac{3}{8} = 9\frac{5}{8}$

$9\frac{5}{8}$ lb

17 $4\frac{1}{3} + 2\frac{5}{6} = 4\frac{2}{6} + 2\frac{5}{6} = 6\frac{7}{6} = 7\frac{1}{6}$

$8 - 7\frac{1}{6} = \frac{5}{6}$

$\frac{5}{6}$ ft

Chapter 4 Addition and Subtraction of Fractions

1 A

2 B

3 C

4 D

5 A

6 $26\frac{2}{3}$

7 $2\frac{9}{40}$

8 $14\frac{15}{16}$

9 $6\frac{43}{60}$

10 $\frac{1}{8}$

11 $\frac{26}{55}$

12 $3\frac{3}{40}$

13 $5\frac{3}{10}$

14 $6\frac{11}{18}$

15 $1\frac{5}{16} + 1\frac{5}{16} = 2\frac{10}{16}$

$1\frac{5}{16} + 2\frac{10}{16} = 3\frac{15}{16}$

$3\frac{15}{16}$ lb

16 $3\frac{8}{9} - 3\frac{1}{5} = 3\frac{40}{45} - 3\frac{9}{45} = \frac{31}{45}$

$3\frac{40}{45} + \frac{31}{45} = 3\frac{71}{45} = 4\frac{26}{45}$

$4\frac{26}{45}$ L

17 $(8 + 5) \div 3 = \frac{13}{3} = 4\frac{1}{3}$

$4\frac{1}{3} - 1\frac{3}{5} = 3\frac{5}{15} - \frac{9}{15}$

$= 2\frac{20}{15} - \frac{9}{15} = 2\frac{11}{15}$

$2\frac{11}{15}$ cups

Continual Assessment 1

1 B

2 A

3 C

4 B

5 D

6 D

7 C

8 B

9 A

10 C

11 10,010,000

12 70,538,011

13 <

14 D, B, A, C

15 2; 16

16 $(10 + 3) \times 2 = 13 \times 2 = 26$
or $2 \times (10 + 3) = 2 \times 13 = 26$

17 10,800

18 86 R5

19 $33\frac{1}{3}$

20 6

21 $84,000 - 4,900 = 79,100$
$79,100 \div 14 = 5,650$
$5,650

Continual Assessment 1

22

Bottle 〔236〕

Bucket

Pitcher

2 units $\longrightarrow$ 236

1 unit $\longrightarrow$ 236 ÷ 2 = 118

17 units $\longrightarrow$ 17 × 118 = 2,006

2,006 mL

---

23 468 ÷ 6 = 78

78 − 65 = 13

13 bags

---

24 (892 × 12) + (168 × 6)

10,704 + 1,008 = 11,712

11,712 eggs

---

25 $2\frac{1}{5} + \frac{1}{2} = 2\frac{2}{10} + \frac{5}{10} = 2\frac{7}{10}$

$2\frac{1}{5} + 2\frac{7}{10} = 2\frac{2}{10} + 2\frac{7}{10} = 4\frac{9}{10}$

$4\frac{9}{10}$ mi

Continual Assessment 1

1 D

2 B

3 A

4 C

5 C

6 A

7 B

8 B

9 C

10 D

11 fifty-one million, forty-three thousand, seven hundred eight

12 30,000,000

13 100,000

14

300,170,500	312,370,500
324,570,500	336,770,500
348,970,500	361,170,500

15 5,000; 421

16 $(7 \times 11) + 3$
$= 77 + 3 = 80$
or $3 + (7 \times 11)$
$= 3 + 77 = 80$
80

17 2,979,306

18 989 R 78

19 $12\frac{1}{2}$

20 $1\frac{4}{15}$

Continual Assessment 1

21 $65{,}067 \div 23 = 2{,}829$
$2{,}829 \times 93 = 263{,}097$
263,097 lb

---

22 $8.25 - 2 = 6.25$
$(8.25 \times 2) + (6.25 \times 3) = 35.25$
$35.25

---

23 $4 \times (20 + 15 - 30) = 20$
$20

---

24 $150 - 36 = 114$
$150 + 114 = 264$
$264 \div 6 = 44$
$44 \times 11 = 484$
$484

---

25 $1 - \left(\frac{2}{5} + \frac{1}{3} + \frac{1}{4}\right)$
$= 1 - \left(\frac{24}{60} + \frac{20}{60} + \frac{15}{60}\right) = \frac{1}{60}$
$\frac{1}{60}$

Chapter 5 Multiplication of Fractions

1 A

2 B

3 D

4 B

5 C

6 $\frac{1}{30}$

7 $\frac{3}{4}$

8 $4\frac{1}{2}$

9 $14\frac{1}{7}$

10 $\frac{9}{10}$

11 $25\frac{1}{2}$

12 $\frac{1}{10}$

13 21 limes

14 $7\frac{7}{8}$ mi

15 $\frac{7}{10} \times \frac{1}{2} = \frac{7}{20}$

$\frac{7}{20}$ L

16 $\frac{1}{4} \times 188 = 47$

$188 - 47 + 19 = 160$

160 passengers

17

10 units ⟶ 160

1 unit ⟶ 16

9 units ⟶ 9 × 16 = 144

144 cupcakes

Chapter 5 Multiplication of Fractions

1 C

2 D

3 B

4 C

5 A

6 $\frac{1}{24}$

7 $2\frac{2}{5}$

8 $2\frac{5}{6}$

9 3

10 7

11 $1\frac{3}{16}$

12 6

13 $1\frac{13}{32}$

14 $1 - \frac{1}{6} = \frac{5}{6}$

$\frac{5}{6} \times 150 = 125$

$125

15 $3\frac{1}{2} \times 1\frac{1}{5} = \frac{7}{2} \times \frac{6}{5} = \frac{21}{5} = 4\frac{1}{5}$

$1\frac{1}{5} + 4\frac{1}{5} = 5\frac{2}{5}$

$5\frac{2}{5}$ cups

16 $\frac{3}{8} \times 208 = 78$

$\frac{1}{6} \times 78 = 13$

13 children

17 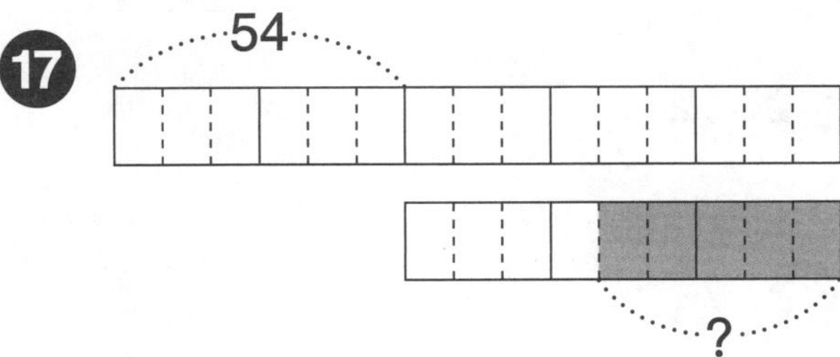

6 units $\longrightarrow$ 54

1 unit $\longrightarrow$ 9

5 units $\longrightarrow$ 45

45 mice

Chapter 6 Division of Fractions

1 C

2 A

3 C

4 B

5 B

6 $\frac{1}{9}$

7 $\frac{3}{50}$

8 12

9 $7\frac{1}{2}$

10 $\frac{6}{7}$

11 $\frac{1}{12}$

12 4

13 $\frac{1}{3}$

14 $\frac{9}{50}$

15 $\frac{7}{10} \div 14 = \frac{1}{20}$

$3 \times \frac{1}{20} = \frac{3}{20}$

$\frac{3}{20}$ kg

16 $\frac{2}{5} \div 6 = \frac{1}{15}$

$2 \times \frac{1}{15} = \frac{2}{15}$

$\frac{2}{15}$

17 $\frac{1}{4} \times 24 = 6$

$6 \div \frac{3}{8} = 16$

or

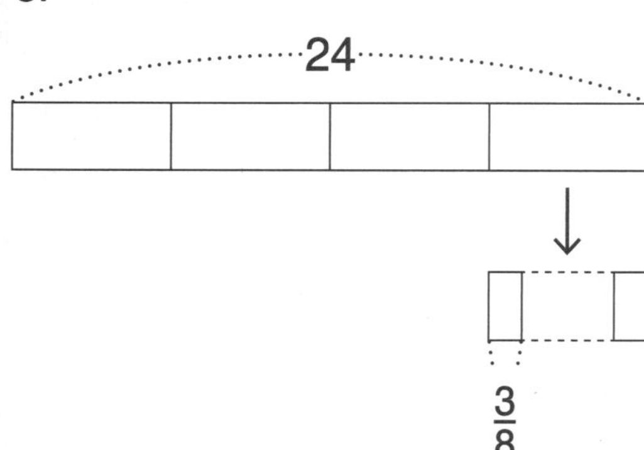

4 units ⟶ 24

1 unit ⟶ 6

$6 \div \frac{3}{8} = 16$

16 bottles

Chapter 6 Division of Fractions

1 B

2 A

3 B

4 D

5 C

6 $\frac{1}{100}$

7 $\frac{11}{40}$

8 360

9 $16\frac{1}{2}$

10 $6\frac{1}{4}$

11 $\frac{7}{45}$

12 $3\frac{1}{12}$

13 9

14 $\frac{11}{48}$ ft

15 $9\frac{1}{4} - \frac{1}{4} = 9$

$9 \div \frac{6}{10} = 9 \times \frac{10}{6} = 15$

15 bags

16 $1 - \frac{1}{4} = \frac{3}{4}$

$\frac{3}{4} \div 9 = \frac{3}{4} \times \frac{1}{9} = \frac{1}{12}$

$\frac{1}{12} \times 5 = \frac{5}{12}$

$\frac{5}{12}$

17 $\frac{3}{4}$ lb $\longrightarrow$ \$6
1 lb $\longrightarrow$ \$6 $\div \frac{3}{4} = 8
$3\frac{1}{2}$ lb $\longrightarrow$ \$8 $\times 3\frac{1}{2} = $8 \times \frac{7}{2} = 28
$1\frac{5}{8}$ lb $\longrightarrow$ \$8 $\times \frac{13}{8} = 13
\$28 − \$13 = \$15

or

$\frac{3}{4}$ lb $\longrightarrow$ \$6
1 lb $\longrightarrow$ \$6 $\div \frac{3}{4} = 8
$3\frac{1}{2} - 1\frac{5}{8} = 3\frac{4}{8} - 1\frac{5}{8} = 2\frac{12}{8} - 1\frac{5}{8} = 1\frac{7}{8}$
$1\frac{7}{8}$ lb $\longrightarrow$ \$8 $\times \frac{15}{8} = 15
\$15

Chapter 7 Measurement

1 A

2 C

3 C

4 D

5 B

6 5 lb 4 oz

7 7,200 m

8 $5\frac{4}{9}$ in²

9 15 cm²

10 QR

11 5 cm

12 52 cm²

13 $27\frac{1}{2}$ cm²

14 35 cm²

15 $\frac{3}{4}$ h = 45 min

2:00 + 45 min ⟶ 2:45

$1\frac{1}{12}$ h = 1 h 5 min

2:45 + 1 h 5 min ⟶ 3:50

3:50 p.m.

16 $10 - 1 - 1\frac{1}{3} = 7\frac{2}{3}$

$6 \times 7\frac{2}{3} = 46$

$6 + 2\frac{1}{4} + 2 = 10\frac{1}{4}$

$10\frac{1}{4} \times 10 = \frac{41}{4} \times 10$

$= \frac{410}{4} = \frac{205}{2} = 102\frac{1}{2}$

$102\frac{1}{2} - 46 = 56\frac{1}{2}$

$56\frac{1}{2}$ m²

17 $\frac{1}{2} \times 7 \times 3 = \frac{21}{2} = 10\frac{1}{2}$

1 ft 3 in = 15 in

15 × 7 = 105

$105 - 10\frac{1}{2} = 94\frac{1}{2}$

$94\frac{1}{2}$ in²

Chapter 7 Measurement

1 D

2 B

3 C

4 B

5 A

6 10 m 5 cm

7 15 qt

8 $6\frac{1}{3}$ in²

9 EJ

10 20 cm²

11 60 cm²

12 48 cm²

13 330 cm²

14 72 cm²

15 $\frac{1}{3} + 2\frac{5}{12} = \frac{4}{12} + 2\frac{5}{12} = 2\frac{9}{12} = 2\frac{3}{4}$

10:30 a.m. $+ 2\frac{3}{4}$ h $\longrightarrow$ 1:15 p.m.

1:15 p.m.

16 $10\frac{1}{2} \times 8 = \frac{21}{2} \times 8 = 84$

$10\frac{1}{2} - 5 = 5\frac{1}{2}$

$\frac{1}{2} \times 8 \times 5\frac{1}{2} = 4 \times \frac{11}{2} = 22$

$\frac{1}{2} \times 5 \times 2\frac{1}{3} = \frac{5}{2} \times \frac{7}{3} = \frac{35}{6} = 5\frac{5}{6}$

$84 - 22 - 5\frac{5}{6} = 56\frac{1}{6}$

$56\frac{1}{6}$ in²

17 2 cm = 20 mm

$\frac{1}{2} \times 24 \times 20 = 240$ mm²

$2\frac{1}{2}$ cm = 25 mm

$\frac{240}{25} = \frac{48}{5} = 9\frac{3}{5}$ mm

$25 + 25 + 9\frac{3}{5} + 9\frac{3}{5} = 69\frac{1}{5}$

$69\frac{1}{5}$ mm

Chapter 8 Volume of Solid Figures

1 B

2 A

3 B

4 D

5 C

6 1,620 cm³

7 7 cm

8 25 in

9 (a) 575

 (b) 1; 950

10 8 × 6 × 7 = 336
8 × 3 × 9 = 216
336 + 216 = 552
552 cm³

11 19 − 17 = 2
30 × 20 × 2 = 1,200
1,200 cm³

12 10 × 7 × 8 = 560
6 × 3 × 8 = 144
560 − 144 = 416
416 cm³

13 240 ÷ 10 ÷ 8 = 3
3 + 6 = 9
9 in

Chapter 8 Volume of Solid Figures

1 D

2 C

3 B

4 A

5 B

6 1,089 cm³

7 26 cm

8 16 cm

9 (a) 3,005

(b) 9; 50

10 $\frac{2}{3} \longrightarrow 10$ cm
whole height $\longrightarrow 10 \times \frac{3}{2} = 15$ cm
$25 \times 8 \times 15 = 3,000$
3 L

11 decrease in water height
$= 14 - 8 = 6$
$32 \times 20 \times 6 = 3,840$
3,840 cm³ = 3 L 840 mL
3 L 840 mL

12 $13 \times 6 \times 10 = 780$
$7 \times 7 \times 10 = 490$
$4 \times 18 \times 10 = 720$
$780 + 490 + 720 = 1,990$
1,990 cm³

13 water height in tank = 4 in
total volume of balls
$= 60 \times 3 = 180$
$180 \div 10 \div 9 = 2$
new water height = 2 + 4 = 6
6 in

Continual Assessment 2

1 B

2 C

3 C

4 A

5 D

6 C

7 A

8 B

9 D

10 C

11 630

12 505,000

13 2

14 30 R 4 or $30\frac{2}{13}$

15 25

16 $3\frac{7}{15}$

17 $19

18 72 cm²

19 $4 \times 5 = 20$
$6 \times \frac{5}{3} = 10$
30 cm³

20 Need another 4 cm of water to fill the tank.
$25 \times 10 \times 4 = 1{,}000$ cm³ $= 1$ L

Continual Assessment 2

21

1 unit ⟶ 15

4 units ⟶ 4 × 15 = 60

78 − 60 = 18

18 ÷ 2 = 9

9 stickers

---

22 $12 \div \frac{2}{5} = 30$

30 × 6 = 180

$180

---

23 10 ft = 120 in

$\frac{1}{3}$ ft = 4 in

120 + 4 = 124

8 × 12 = 96

124 × 96 = 11,904

11,904 in²

24

14 units ⟶ 1,680

1 unit ⟶ 120

3 units ⟶ 360

360 children

25 13 × 3 × 10 = 390

13 − 3 − 3 = 7

7 × 5 × 10 = 350

390 + 350 = 740

740 cm³

Continual Assessment 2

1 C

2 A

3 C

4 D

5 B

6 A

7 B

8 C

9 A

10 C

11 73,820

12 <

13 345; 3

14 11,730

15 $2\frac{1}{4}$

16 $15\frac{1}{30}$

17 10 cookies

18 22 ounces

19 60 cm²

20 55 cm²

21 192 + 64 − 28 = 228
228 ÷ 4 = 57
57 × 6 = 342
$342

22 $18 \times \frac{2}{3} = 12$
12 ÷ 6 = 2
$\frac{2}{18} = \frac{1}{9}$
$\frac{1}{9}$

Continual Assessment 2

23

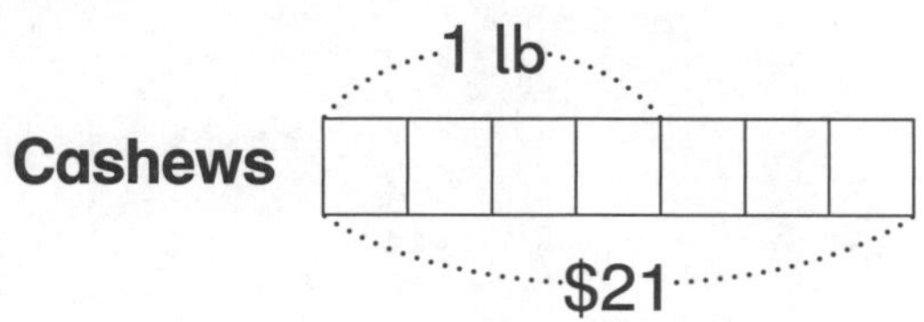

1 unit ⟶ $3

3 units ⟶ $9

7 units ⟶ $21

1 unit ⟶ $3

4 units ⟶ $12

12 − 9 = 3

$3

or

$\frac{1}{3}$ lb ⟶ $3

1 lb ⟶ $3 ÷ $\frac{1}{3}$ = $9

$\frac{7}{4}$ lb ⟶ $21

1 lb ⟶ $21 ÷ $\frac{7}{4}$ = $12

12 − 9 = 3

$3

---

24 7 × 6 × 10 × 2 = 840

6 + 7 + 6 = 19

19 × 8 × 10 = 1,520

840 + 1,520 = 2,360

2,360 cm³

25 Increase in water level:

11 − 9 = 2

Increase in volume:

420 × 2 = 840

Volume of cuboid:

7 × 5 × 2 = 70

Number of cuboids:

840 ÷ 70 = 12

12 cuboids

or

Increase in water level:

11 − 9 = 2

Number of cuboids:

$$\frac{420 \times 2}{7 \times 5 \times 2} = \frac{420}{7 \times 5} = 12$$